Managing the Unknown

Managing the Unknown

A New Approach to Managing High Uncertainty and Risk in Projects

Christoph H. Loch
Arnoud De Meyer
Michael T. Pich

John Wiley & Sons, Inc.

This book is printed on acid-free paper. ♾

Published by John Wiley & Sons, Inc., Hoboken, New Jersey
Published simultaneously in Canada

For general information about our other products and services, please contact our Customer Care Department within the United States at (800) 762-2974, outside the United States at (317) 572-3993 or fax (317) 572-4002.

Wiley also publishes its books in a variety of electronic formats. Some content that appears in print may not be available in electronic books. For more information about Wiley products, visit our web site at www.wiley.com.

Library of Congress Cataloging-in-Publication Data:

Loch, C. (Christoph)
    Managing the unknown: a new approach to managing high uncertainty and risk in projects / Christoph H. Loch, Arnoud De Meyer, Michael T. Pich.
      p. cm.
    Includes bibliographical references and index.
    ISBN-10: 0-471-69305-7 (cloth)
    ISBN-13: 978-0-471-69305-5 (cloth)
    1. Project management. 2. Risk management. I. Meyer, Arnoud De.
II. Pich, Michael T. III. Title.
HD69.P75L62 2006
658.4'04--dc22

2005019102

Printed in the United States of America

10 9 8 7 6 5 4 3 2 1

Contents

Foreword

Project management, as an important part of our disciplinary back-ground, has been close to our interests for a long time and has been part of our professional experience as consultants, program directors, and, in one case, dean at INSEAD. The research on which this book is based began in 1996, when we became interested in project risk management and, in particular, methods that might help project managers to deal with novel projects.

We first explored real options and contingent decision-making in projects, a "new" method from the finance discipline that was widely discussed in the mid-1990s. This turned out to be a dead end, as the powerful analytical methods from finance did not sufficiently carry over to project management, an environment where much less information is available than in financial markets. In project management, these methods were wonderful in theory but required too many assumptions to apply in practice.

As we worked with project managers and wrote cases about the projects we analyzed, we began to realize that our concerns with managing novel projects were not so much with the existing project risk management (PRM) methods per se—as these methods were quite well developed over many years—but with the very concept of risk itself. We felt that the concept of risk, as discussed in PRM, was not entirely appropriate for novel projects. The first practical lessons we learned from these project managers were summarized in a *Sloan Management Review* paper in 2001. But we also understood that we needed a conceptual framework to help us understand how seasoned project managers of novel projects dealt with uncertainty and risk, and how a project team might in principle deal with it.

We realized that a key reason many projects fail is because organizations do not recognize the fundamental difference between project novelty and project risk. This thinking led to a publication in *Management*

Science (2002), in which we examined the key difference between project risk and project novelty. We defined two critical characteristics of novel projects: unforeseeable uncertainty and complexity. We also identified two fundamental approaches that were, in principle, possible in the presence of unforeseeable uncertainty and complexity. We referred to these new approaches as selectionism and learning.

These theoretical considerations gave us a "roadmap," a lens through which we could gather evidence and examine the projects with which we worked. We discovered that, while selectionism and learning had, of course, been discussed before, a complete approach to managing novel projects was missing. How should organizations and project management teams approach novel projects? How do they diagnose the level of project novelty, how do they build the organizational capabilities to deal with novel projects, and finally, how do they go about managing these projects?

This, then, is the new thinking and the contribution of this book. Selectionism and learning can be viewed as a powerful arsenal of weapons in the face of high novelty and unforeseeable uncertainty in projects, an arsenal *in addition* to established project risk management methods. The book explores the nature of unforeseeable uncertainty, of "unk unks" and explains the concept of selectionism and learning. Then, most of the book tries to develop operational and actionable methods and tools that managers can use to evaluate selectionism and learning and to put them in practice. This book may be interesting for academics, but we are addressing an audience of practicing managers, and we hope they find it useful.

We are indebted to the many project managers with whom we have worked and from whom we have learned—in our case research and in our management seminars at INSEAD, where the participants learn from us, but where we also learn invaluable insights from them.

We also thank our colleagues, notably Christian Terwiesch and Svenja Sommer, the collaboration with whom has helped us think through many issues. We are indebted to Jean Cropper, who helped us turn our clumsy language into readable English.

If this book makes a small contribution to the project management profession, we feel proud, and the tremendous effort was worthwhile.

Fontainebleau and Singapore,
May 2005

Christoph Loch
Arnoud De Meyer
Michael T. Pich

Introduction

Project Management and Project Risk Management (PRM)

Nearly all managers engage in project management. The rapid change in the competitive climate, the internationalization of the economic environment, and the rise in customer power have rendered business activities less repetitive. Today, all managerial tasks contain elements of project management. In older textbooks, project management was often identified in the context of industrial activities, such as construction, product development, or the introduction of new technologies. The most important challenge in these projects was often to handle myriad tasks and relationships, and the focus of project management methods was on complexity. But today's projects are much broader, encompassing, among other activities, internal or external consulting, launching market campaigns, developing software, or implementing mergers or acquisitions. These new projects are often less complex than the traditional ones, but they also often break new ground and are therefore confronted with more uncertainty. This book is about highly novel projects that have to cope with a high level of uncertainty.

What is a project? A project can be defined as a sequence of activities undertaken to accomplish a temporary endeavor (with a defined completion date) to create a unique product or service.[1] Projects represent an organizational tool to respond to risks. They are made up of temporary structures, they use flexible methods, and they are dismantled after the work is done. Each project is unique in some respect: It can be a new project, meaning that the tasks have not been done before, or there can be spatial separation, indicating that the resources are assembled at a separate place. Each project is thus managed slightly differently from other projects or ongoing repetitive processes. In short, projects are temporary structures and management methods that allow companies to be flexible and to respond to risks.

The ability of an organization to use temporary structures to flexibly respond to risks has gained in relevance in many business contexts. While projects and project management methods have historically been relegated to the R&D, infrastructure, IT, and defense-related industries, projects and the language of projects have now become ubiquitous in a wide variety of industries, organizations, and functional areas. Projects are now seen as an important strategy to manage change in organizations. This reflects an increase in uncertainty and velocity in many industries, which has prompted prediction in the popular press that "the old organizations and career tracks will be obliterated in the future."[2]

This has brought about a critical look at projects, starting with empirical evidence of the prevalence of project failure. The Project Management Institute (PMI) states that over 70 percent of projects in large organizations fail to meet their stated objectives. A study of large engineering projects by Miller and Lessard found that only 45 percent of these projects met most of their stated objectives, while 19 percent met only a subset of their objectives, 16 percent had to be significantly restructured, and fully 20 percent were abandoned. The picture is no better for startup venture projects. Before the 1997–2001 bubble, nearly 50 percent of the value created by venture capital firms was a result of the 6.8 percent of their investments that actually achieved a return of 10 times or more. Battery Ventures reported that of 153 companies it backed, 25 went public, while 43 were acquired—and 25 companies went out of business.[3]

Many reasons are given for this rather poor record of project success, ranging from technical or market uncertainty, poor leadership, and multiple stakeholders with differing interests. Too often, these failures to meet stated project objectives are interpreted by organizations as failures on the part of the project management team, negatively affecting promotions and bonuses. While each of these reasons may be present in various projects, we will argue that the main reason for project failure is that organizations do not recognize the fundamental difference between project novelty and project risk.

An understanding of project risk and project risk management (PRM) has been around for a very long time. Project risk management has become an integral part of project management and is an established, formalized, and widely used project management method. This is reflected by a steady stream of technical books on the topic.[4]

Project risk is often defined as the product of the probability of an event's occurrence and the extent of its impact. In other words, what is the probability that an event will occur, and what is the extent of impact of this event on the project, as planned, if it *does* occur? PRM techniques have thus been developed to help organizations to *identify* possible uncertain events, *assess* their potential impact (positive or negative) on the project plan, *prioritize* these events for action, and *develop* preventive, mitigating, and contingent actions in response to these foreseen events. PRM even goes so far as to foresee that not all events can, or will, be predicted, and that organizations must be prepared to quickly improvise a response to this "residual risk" to get the project back on track. Issues of project leadership, stakeholder management, and supplier relationships are all dealt with in an attempt to create a project infrastructure and environment that can flexibly respond to events in order to bring the project back under control, should events, both foreseen and unforeseen, arise.

But project novelty is not the same as project risk, and novel projects pose fundamentally different challenges than do risky projects. Novel projects, by definition, cannot be planned. In novel projects, there are either too many "unknown unknowns" (unk unks) or there is too much complexity, or both. There is no "real" project plan around which to apply standard PRM techniques. Contingency planning makes little sense, as the major risks are unknown. In this context, complexity means that even if events could be foreseen, the interaction of events is so complex that contingency plans would be impossible to draw up. Novel projects pose unforeseeable uncertainty and are complex in nature, and this is necessarily so because only such projects offer sustainable rents. Simple projects are easily copied and their rents competed away. Increasing industry dynamics and sophistication force project managers to push the envelope in seeking new markets and new technologies.

The fundamental logic of the PRM mind-set does not address novel projects. PRM preplans contingencies and flexibility around a project plan and then "triggers" them as events unfold. This approach works fine as long as all risks are identified, or as long as residual risks are simply events that temporarily take the project off its planned course. The basic backbone of PRM is that there is a real project plan, one that can be implemented, albeit with some contingent or mitigating actions thrown in. When unforeseen events arise, the project team can improvise its way back to the basic project plan because it is never too far away from it.

Serious problems arise when the PRM mind-set and toolbox are applied to novel projects. Because one cannot have a project without a plan, a project plan will inevitably be drawn up for the novel project. As the project is "novel," rigorous PRM techniques will be applied to ensure that all potential risks are identified, assessed, and prioritized, and that preventive, mitigating, and contingent actions are developed. Solid, experienced project leadership will be allocated to the project, supplier relationships will be considered, and good project governance will be installed, all around the idea that when unforeseen events arise, the project team, its suppliers, and its stakeholders will all be prepared to improvise a "solution" to these unforeseen events and bring the project back on plan. But, of course, there *is* no project plan. The plan is but a starting point, an illusion, a simple sketch. The basic backbone of PRM, the project plan, does not really exist.

This creates tremendous problems for the project team. All the PRM efforts provide a level of comfort to all the parties involved that the project, as planned, is feasible. The team is convinced that their job is to implement the plan, without properly appreciating the challenges that lie ahead. As unforeseen events arise, they try to improvise their way "back" to the project plan. But the project, as planned, may not have been feasible, and this leads to two common stages of behavior. In the first stage, the project team makes promises that the project will be back on track after some degree of effort. The improvisational effort is undertaken, and the project is still not back on track. This cycle continues for a while until the second-stage behavior manifests itself: The team suddenly realizes that the project plan was simply stakes in a desert, and they wonder how they got out there themselves. All their effort had been concentrated on following the stakes in the ground, instead of doing what they really should have done: discovering where the stakes should be going.

Operational Risk Management and PRM

Before we explain how this book starts from PRM and then turns to managing unforeseeable uncertainty, we must distinguish project risk from the term "(enterprise) risk management," which has been prominently discussed in the business press since the banking and accounting fraud scandals of the late 1990s. In both finance and strategy literatures, one can find other definitions of operational risk than the one we will use with respect to projects.

With the Asian crisis and the bursting of the financial markets bubble in 2000, attention has focused on market risks, credit risks, liquidity risks, and legal/regulatory risks. Then, in the wake of corporate governance and fraud scandals, such as Enron, Tyco, Ahold, Parmalat, and Arthur Andersen, the interest conflicts between stock market analysts and investment bankers within large, universal banks, and many others turned attention to what the financial community began to call "operational risk": the risk of direct or indirect loss resulting from inadequate or failed internal processes, people, and systems or from external events, the latter also being called "operational strategic risk."[5]

In the context of business strategy, *strategic risk* is defined as "an unexpected event or set of conditions that significantly reduces the ability of managers to implement their intended business strategy."[6] Strategic risk comprises operations risk, asset impairment risk, and competitive risk. Operations risk results from the consequences of a breakdown in a core operating, manufacturing, or processing capability. Any operational error that impedes the flow of high-quality products or services has the potential to expose the firm to loss and liability. At the aggregate business level, there have also been recent attempts to cover property, product liability, employee crime, and exchange risks with integrated insurance policies. Overall, however, operational risk is not as well understood as market or credit risk (in particular, mathematical models are not as fully developed), and many financial institutions manage operational risk on an ad hoc basis.[7]

There are important differences between "operational risk management," as seen by strategy and financial services, and PRM. PRM is more focused than operational risk management because it looks at projects only.

At the same time, it is more general because it considers any possible event that may change the execution of the project plan or the plan itself, over the course of the project. PRM is more operational than operational risk management because it is directly concerned with the work of the project team—namely, *what* they need to do and *how* flexible they must be in responding to stochastic events. Finally, the notion of "risk" is more general in PRM: In contrast to operational risk, project risk may represent not only a *downside* (such as adverse events that threaten the project) but also an *opportunity*, or *upside* (such as an unexpected application of a project result). PRM should include both guarding against unpleasant surprises *and* taking advantage of opportunities.

Contribution and Plan of the Book

In this book, we make two main contributions. First, we begin by discussing project risk management (PRM), because novel projects have a lot to do with risk and PRM is the best toolbox available to handle them. Then we explore the implications of *unforeseeable uncertainty* in novel projects, and we demonstrate that they render PRM methods insufficient and require *fundamental modifications in the logic of project management itself*. Thus, we offer a different way of looking at the problem of managing novel projects: Rather than working harder, straining to consider more possibilities in planning and formal risk management, we propose that one must recognize that project novelty fundamentally renders planning approaches inadequate. In such cases, project managers must make use of two fundamentally different approaches: *learning*, or a flexible adjustment of the project approach as one learns more about the project, its environment, and their interactions (as opposed to a contingent approach that utilizes planned "trigger points"), and *selectionism*, or pursuing multiple approaches, independently of one another, and picking the best one ex post.

Second, these approaches must be managed differently from each other and from the traditional planning approach; they require different *project mind-sets*, different *project infrastructures,* and different *supplier and stakeholder relationships*. Project sponsors must be aware of these differences if they are to help build the organizational capabilities critical to managing novel projects and to properly support these novel projects within a possibly hostile organization.

While the important events and courses of actions cannot be foreseen in novel and complex projects, the project team, more often than not, has a good feel for the complexity of the project and for the limitations of its own knowledge, thus preventing it from falling victim to unk unks. The team can, at the outset, put in place the competencies, infrastructure, and relationships appropriate to the challenge at hand. This is what we propose in this book, and for which we offer tools. These tools are not as quantitative and precise as traditional PRM tools, reflecting the more difficult nature of the challenges that confront novel projects. Yet we set out to convince the reader that they can make an enormous difference to the project.

In Part I of this book, we present two contrasting examples of current PRM practice and then propose an extended view of PRM. Chapter 1 describes a PRM *best practice* example, while Chapter 2 illustrates the limitations of PRM practice in novel projects. Chapter 3 proposes a broader view of PRM and demonstrates that, in the presence of unk unks, project management itself must follow an extended logic. We distinguish fundamental sources of uncertainty and expand the PRM toolbox in order to address uncertainty from all these sources.

Part II is the conceptual heart of the book, where we explain what a project can do in the face of unforeseeable uncertainty, not only in terms of PRM but by changing the project management approach itself. Chapter 4 offers a diagnosis tool for recognizing the type of uncertainty at the outset of a project and outlines the two fundamental approaches to unk unks, selectionism and trial-and-error learning. Chapters 5 and 6 explain and give examples of learning projects and selectionist projects, respectively. Chapter 7 develops tools for choosing between the two and for combining them and the strengths of each.

Part III develops tools for managing learning and selectionist projects facing unforeseeable uncertainty on three fundamental dimensions. Chapter 8 defines the culture and mind-set that are necessary to be successful in the face of unk unks, a mind-set of experimentation and learning rather than of executing planned tasks and achieving established targets. Without such a mind-set, tools will not be successful. We also discuss what this means for the profile of the members of the project team. Chapter 9 describes the required project infrastructure, the systems that need to be put in place for planning, monitoring, coordinating, evaluating, and exchanging information. These management systems must vary according to what fundamental project management approach is chosen.

Chapter 10 discusses the collaboration with partners, an increasingly important way of organizing large projects with wide-ranging technologies. Managing partners becomes more challenging in the presence of unk unks, as unexpected surprises tend to unbalance the interests of the players involved (even if they were perfectly aligned at the outset), which produces an often irresistible temptation to behave opportunistically in order to safeguard one's own interests. The chapter describes how flexible project governance with limits of acceptable behavior can be achieved using a collection of contracts, informal agreements, mutual stakes and ownership, repeated interactions (into the future), experience, and personal relationships that are necessary in order to withstand the pressures of unexpected surprises.

Chapter 11 addresses the project stakeholders. Stakeholders are parties that are not formally involved in the project (as opposed to the partners in Chapter 10) but are affected by the project, interested in it, and can influence it through formal means and informal means and the application of power. The presence of unk unks renders an appeal to rational interests insufficient—interests may change along with the content of the project over time, as unk unks emerge.

Finally, Part IV elevates the discussion from project management to senior management. While project management must execute the project and deal with the unk unks, senior management shapes the project's environment and is responsible for enabling the project team to respond to unexpected events. Thus, senior management has a significant responsibility in making novel projects successful. Chapter 12 outlines three areas of responsibility that senior management should keep in mind.

This book is meant as a resource for project teams that have to deal with novel projects. We offer a number of tools and ideas for your inspiration, but even then, we realize that this does not make managing novel projects trivial. Dealing with unknown unknowns is inevitably uncomfortable and dangerous. We hope this book provides some guidance or red thread in the chaos of dealing with the many unexpected and hard-to-interpret events that will inevitably arise in novel projects.

Endnotes

1. This is a standard definition; see, for example, Meredith and Mantel 2003, p. 8.

2. An increase in velocity, and thus the need for project management, has been observed by a number of researchers, for example, Kloppenborg and Opfer 2002, Pinto 2002, and Kerzner 2003. The conclusion that career tracks will be obliterated can be found, for example, in Stewart 1995. This latter conclusion is premature.

3. See Sahlman 1990. He reports that of the 383 companies studied between 1969 and 1985, 34.4 percent experienced a partial or total loss of invested capital, 49.8 percent returned between 0 and 5x, and 9.8 percent returned between 5 and 10x. While the Internet bubble for a while promised spectacular returns, VCs are now back to the times before 1997 and have actually begun to focus more on incremental projects.

4. For example, Wideman 1992, PMI 1996, Chapman and Ward 1997, the Department of Defense (2001), or Smith and Merritt 2002.

5. The focus on market, credit, liquidity, and legal risks is observed in Crouhy *et al.* 2000, p. 35. The term "operational strategic risk" is coined ibid, p. 479. The focus on inadequate or failed internal processes, people, and systems or on external events is described in Basel Committee Publications 1998, and Harmantzis 2003.

6. See Simons 1999.

7. Integrated insurance policies are described in Meulbroek 2001. The diagnosis that operational risk is often managed ad hoc is in Crouhy et al. 2000, p. 484.

A New Look at Project Risk Management

Imagine you are planning a climbing expedition up the Matterhorn, one of the most spectacular mountains in the Alps. As a project management expert, you produce a detailed plan and specify routes, expected distance travel times, budgets for equipment, shelter and food, and so on. In addition, you have to worry about what might go wrong: A storm may move in, for example. For such an eventuality, you need to build buffers into the plan: extra time and/or extra equipment (perhaps an emergency tent or ice picks). You also need to plan decision points; for example, if the storm moves in before noon, we turn around, and if it

Copyright hak@tenit..com.pl

catches us at 4:00 P.M., we take refuge in an emergency shelter. This exercise of antici-pating risks is the essence of project risk management (PRM).

Project risk management can be defined as the "art and science of identifying and responding to project risk throughout the life of a project and in the best interest of its objectives." PRM extends project planning by identifying, appraising, and manag-ing project-related risks. Risk, in turn, is defined as "the implications of the existence of significant uncertainty about the level of project performance achievable" and is seen as having two components: first, the probability of occurrence, and second, the consequences or impacts of occurrence.[1]

PRM has become an established, formalized, and widely used project management method.[2] This method offers a powerful set of tools that help companies to keep downside risks under control and to take advantage of upside opportunities. In some industries, such as engineering, construction, or pharmaceuticals, anticipating and man-aging downside risks is essential to remain in business. In other industries, the ability to seize opportunities can greatly enhance profitability. For example, the manager of one power generation engineering company told us, "Thinking proactively through risks enables us to fill the 'white spaces' in our contracts to our advantage. We proactively interpret undefined events in our favor. 'User training costs money? Well, that was not specified, so it's clear that *you* must pay it.' This protects us from the customer interpret-ing the event in *his* favor, and sometimes, we even manage to seize an opportunity and sell it to the client for additional profit."

PRM, then, is concerned with *achieving the stated project goals* in spite of risks (see Smith and Merritt 2002), although it ideally includes influencing the "base plan" and even the project design, and revising the targets when necessary.[3] While the details differ, authors agree that PRM consists of four conceptual steps (see Figure I.1): *Identify* risks beforehand; classify and *prioritize* them according to probability and impact; *manage* them with a collection of preventive, mitigating, and contingent actions that are triggered by risk occurrence; and *embed* these actions into a system of documentation and *knowledge transfer* to other projects.

In Part I of this book, we present two examples of project risk management. The first example, the PCNet project, describes one of the many IT integration projects under-taken as part of the takeover of RBD, Inc. by the diversified resources company Metal Resources Co. from July 2001 to September 2002. We consider this to be an excellent example of how solid application of PRM techniques in the appropriate project envi-ronment can produce good results.

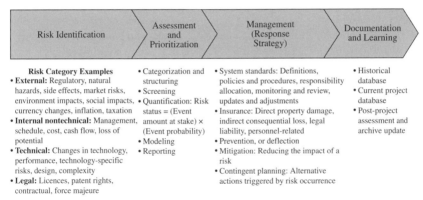

Risk Identification	Assessment and Prioritization	Management (Response Strategy)	Documentation and Learning
Risk Category Examples • **External:** Regulatory, natural hazards, side effects, market risks, environment impacts, social impacts, currency changes, inflation, taxation • **Internal nontechnical:** Management, schedule, cost, cash flow, loss of potential • **Technical:** Changes in technology, performance, technology-specific risks, design, complexity • **Legal:** Licences, patent rights, contractual, force majeure	• Categorization and structuring • Screening • Quantification: Risk status = (Event amount at stake) × (Event probability) • Modeling • Reporting	• System standards: Definitions, policies and procedures, responsibility allocation, monitoring and review, updates and adjustments • Insurance: Direct property damage, indirect consequential loss, legal liability, personnel-related • Prevention, or deflection • Mitigation: Reducing the impact of a risk • Contingent planning: Alternative actions triggered by risk occurrence	• Historical database • Current project database • Post-project assessment and archive update

Figure I.1   Conceptual steps of the PRM process

The second example, the Circored project, describes the design and construction of a plant in Trinidad to produce direct-reduced iron (DRI), as part of a joint venture between Cleveland Cliffs, one of the largest iron ore suppliers to blast furnace integrated steelmakers in the United States, Lurgi Metallurgie GmbH, a metallurgical process engineering company, and LTV Steel, who wanted to use DRI in a mini-mill they were building in Alabama. We consider this to be an excellent example of what happens when standard PRM techniques are applied to novel, first-of-a-kind projects. As is often seen in such cases, the project ran into unexpected problems that delayed completion for several months, it ran over budget, and it was eventually blindsided by an unexpected turn in the market.

In Chapter 3, we draw the lessons from the two examples and characterize the *types of uncertainty* that require PRM. Based on this classification, we outline under what circumstances which of the various methods of PRM are appropriate. In addition, we extend the PRM toolbox by discussing additional methods that are relevant but have not been presented in the same context. Control-and-fast-response is a method of dealing with high task complexity, and project contracts are used to coordinate multiple stakeholders in the presence of relationship complexity. Finally, we introduce two approaches to unforeseeable uncertainty, both of which can extend PRM: trial-and-error learning, or the repeated redefinition of the project over time, and selectionism, or the use of parallel candidate trials, the best of which is chosen ex post. These will be further discussed in Part II of the book.

Endnotes

1. The definition of PRM can be found in Chapman and Ward 1997, p. 10; see also Wideman 1992. The definition of risk is in Chapman and Ward 1997, p. 7. The components of risk are described in DoD 2001, p. 5.

2. See, for example, Wideman 1992, PMI Standards Committee 1996, Chapman and Ward 1997, Council of Standards Australia 1999, DoD 2001, or Smith and Merritt 2002.

3. See Chapman and Ward 1997, p. 10.

PRM Best Practice: The PCNet Project[1]

1.1 Background

In 2002, the first-tier diversified resource company, Metal Resources Co., headquartered in Austin, Texas, announced a cash offer for the Winnipeg-based metals company, RBD, Inc. The offer was recommended by the RBD board to its shareholders and then swiftly executed. The combined companies formed the second largest mining and resources company in the world. In 2004, Metal Resources Co. had activities in 28 countries, with $29 billion in sales, 40,000 employees, and leadership positions in aluminum, nickel, copper, uranium, gold, carbon steel metals, diamonds, manganese, and various specialty metals used in steel production.

The acquisition execution placed heavy emphasis on synergies, that is, on gross annual cost savings. Five top-level integration areas were put in place to capture the savings: IT infrastructure, HR systems and processes, financial systems and processes, operational integration, and organizational integration. The total synergy target amounted to $1.9 billion in annual gross operating cost savings.

One of the important merger activities was a consolidation of the IT systems of the two companies, a huge undertaking involving 900 IT employees throughout the combined company. Not only had the IT integration achieved its own target of $210 million in savings, but it also critically enabled the other merger areas by providing a transparent, integrated, and reliable application platform. Max Schmeling, the enterprise CIO, was responsible for executing the IT integration.

A pre-integration team planned the integration project in detail over a period of nine months, up to October 2002. The team started from an overall target ("get $210 million in annual savings by consolidating the IT structure") and successively broke this target down to more and more detailed tasks. Within each consolidation area, large projects were defined, then subprojects, and then detailed tasks that could be assigned. In total, 110 projects were to be executed in parallel by project managers and sub-project managers, supported by a central project office. The projects would have to be carefully coordinated, as some of them served as enablers for others, and all competed for the same scarce staff time.

In September 2002, Max Schmeling pulled his direct reports and operating company CIOs into one room (about a dozen people). He presented them with the work breakdown structure that the planning team had produced. He said, "Nobody leaves this room before every one of the 110 savings opportunities has a name on it." The first outcome of this two-day marathon meeting was a project structure that clearly assigned project accountabilities, as well as a corporate sponsor for each project, to help them drive change through the organization. The key projects within the IT integration were "General" (mainframe decommissioning, Unix integration, and office consolidation at the various sites of the previous companies), knowledge management (including the consolidation of portals, intranets, yellow pages, instant messaging, collaboration tools, and publishing), the ERP program (moving to an enterprisewide SAP installation encompassing HR systems and financial reporting across both companies), the Web applications center, IT strategy (which was to connect the IT changes to head counts and reengineered processes, and strategic IT sourcing), and the PCNet project, on which our project risk management (PRM) example focuses.

The last piece, not producing bankable savings but critical nonetheless, was the "Time Zero" project: The IT systems were changed while simultaneously running the business. Critical systems, such as e-mails, global address lists, help desks, and all business applications (but not, for example, cross-unit calendar lookup), had to work on day one. Without this minimal functionality, the business damage would have been too great, and resistance would have prevented a successful execution of the migration.

The second outcome of the marathon meeting was a "Gantt Chart from Hell," which filled an entire wall. This was a preliminary plan showing how the 110 projects would be sequenced, and when they would reach their critical milestones and, ultimately, completion. In addition to encompassing a large number of tasks, the planning job was made even harder because the 110 projects had to be coordinated with the other synergy projects and with the activities of the operating companies, who were responsible for carrying out numerous activities.

1.1.1 The PCNet Project

The PCNet project encompassed four network infrastructure migration areas (see Figure 1.1): (1) worldwide standard desktop environment, mostly the exchange of the 40,000 companywide desktops, standardizing on Windows XP desktops (HP) and laptops (IBM); (2) a global communications network, consolidating the corporate network (which had previously been centered on the bottlenecks in California and Texas) around six hubs, with added bandwidth to the rest of the network and further internal redundancy; (3) a standard network server infrastructure using Windows 2000, with a greatly reduced number of network routers; and (4) an enterprise security and directory system, going from multiple directories, security systems, and firewalls down to one each. Multiple directories caused headaches when, for example, executives were reassigned and stopped receiving their e-mails until the IT staff had located the directory in which they had been filed.

The business case called for a total savings net present value (NPV) of $115 million, with a project budget of $149 million. This was based on direct savings from infrastructure costs alone, but the project was the key enabler of the whole merger, not just the IT integration. It enabled additional savings, including cutting 130 different applications in the Finance area alone, and it later made the transition to an enterprisewide ERP system much faster and smoother.

The network integration included the reconciliation of outsourcing decisions that had been made differently in the previous companies. For example, Metal Resources had outsourced mainframe, telecom, and the help desk to EDS, while RBD had outsourced the server environment and the help desk to IBM. To move fast, IT management decided to move the entire package to EDS (the provider who already had the bigger share) and to take some server services back in-house.

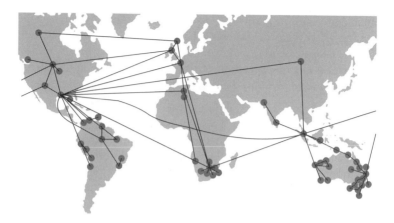

Project objective: Integrate Metal Resources' desktop, network, and server structure, and directory and security services.

- 40,000 desktops migrated to global standard (31,000 from Metal Resources and 9,000 from RBD), investment $89 million, NPV $83 million
- 900 network routers consolidated. Investment $29 million, NPV –$2 million
- Network structure with 500 servers centralized to three hubs with standard servers. Investment $13 million, NPV $20 million
- 10 directories and 7 security systems consolidated to one standard each, Investment $15 million, NPV $14 million

(NPV Assumptions: 4-year savings NPV (after investment), 10% discount rate, tax, and depreciation included)

Figure 1.1   The PCNet project

The PCNet project organization included a project management office, a group for implementation and operations, and a planning group comprising several analysts who compiled business cases and risk analyses, and maintained the tracking tools. Embedded in the master plan for IT integration, the PCNet planning group developed and maintained its own plan and milestones (Figure 1.2). Much of the actual work (such as physically installing routers in basements and desktops in offices) was performed by local staff in the operating companies; the central project organization coordinated, standardized, oversaw time plans, and centrally sourced the standardized components.

1.2 Risk Identification

Risk identification is concerned with recognizing, at the outset, the factors that may make the project plan obsolete or suboptimal. Thus, risk identification is an important part of project planning. It represents a thorough homework exercise that allows the project organization to be prepared when adverse events strike, offering ready-made (rough-cut) solutions, based on which a team can respond quickly.

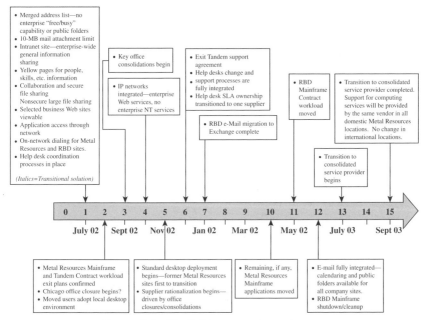

Figure 1.2   The PCNet project key milestones

In the PCNet project, risk planning was a formal part of all project plans. The main risk areas were seen as Operating Company acceptance (they had to perform, and pay for, a lot of the detailed implementation work, and they resented the distraction from their pressing priorities); "staying focused," meaning dealing with too many activities with the same scarce resources; security breaches during the transition; and a change in business climate that would threaten the availability of funds to complete the merger.

This aggregate list rested on many lower-level risk identification efforts, one of which is shown in Figure 1.3, focusing on HR and person-nel retention risks. The list showed where the risk's impact would lie (e.g., in achieving synergies); whether the impact was financial, on the schedule, or on solution quality; and how high the impact would be financially. Finally, the list estimated the risk probability and indicated the impacted parties and the "owner," i.e., the party responsible for responding to the risk.

The lower-level risk lists (such as the one illustrated in Figure 1.3) were produced by the project management office in collaboration with the operating divisions that performed much of the work. The project management office produced a draft list based on experience from previous projects, and the execution teams added risks, based on the detailed tasks, systematically asking what could possibly go wrong. In other words, the risk lists combined knowledge and experience with project-specific planning.

Risk Description	Risk Category	Risk Type	Risk Probab.	Risk Impact	Parties impacted	Risk Owner	Status			Comments
Describe the Risk	Synergy; Day one; Bus. local; Bus. global	Time (T) Financial (F) Quality (Q)	H = >80% M = betw L = < 20%	H = > $5m M = betw L = < $1m	Project Team Impacted by Risk O = Owner I = Impacted	Person responsible for monitoring and reporting	Mitigated?	Open/Closed?	Assigned?	General Notes
			%	$ (one yr.)						
ATTRITION of technology talent could jeopardize exploration, and IT operations and transition	All	T F Q	M	H	I: ALL (operations, exploration, technology, finance, IT/systems, corporate, HR integration, procurement) O: IT/systems	CTO tecnology officers	assigned	open	unmitigated	Retention risk: Medium probability if no adverse event after merger. Operational risk: Spend more on maintaining op co's competences Value risk: Not being focused on few high-value projects, lever 10 times ops risk
RESOURCE RISk— not enough IT staff to handle all projects the functional groups have planned around	All	T F Q	M	H	I, O: IT/systems	IT planning	assigned	open	unmitigated	Good project prioritization and planning needed
CHANGE MGT.— people may not align on priorities (productivity, what things to work on)	All	T F Q	M	M	I, O: IT/systems	CIO and project management.	unassigned	open	unmitigated	Good communication and buy-in is essential.
IT CULTURE— if people do not integrate then wars could arise	All	T F Q	L	M	I, O: IT/systems	New management.	unassigned	open	unmitigated	Good communication and buy-in is essential.

Figure 1.3 HR risk list in the PCNet project

1.3 Risk Assessment and Prioritization

1.3.1. Impact Assessment: The Project Outcome Is a Distribution, Not a Number

After risks have been identified, their impact should be estimated, in order to prioritize them. For the large IT merger projects (from around $10 million), Max Schmeling built on the identified risk factors, and their possible values, to develop a probabilistic distribution of the project outcome. This was done with a scenario business plan with respect to the ultimate success measure, the NPV of savings (see Figure 1.4 for the PCNet project): "With only 10 percent probability, we will fail to deliver savings of at least $90 million; with 50 percent probability, we will not reach $135 million; and with 90 percent probability, we will not be able to deliver as much as $180 million in this project" (in other words, a pessimistic, medium, and optimistic scenario). The scenarios, connected to a value-distribution curve, were called P10, P50, and P90 (a method and terminology that come from mining and oil engineering). In the curve (Figure 1.4), the failure probability increases from left to right as the target increases. The value distribution for the PCNet project rested on similar curves for the four major subprojects.

The project value distribution curve in Figure 1.4 offers two benefits. First, it forces the team and management to acknowledge that the outcome cannot be planned exactly, as a number, but that many outcomes are possible with varying degrees of probability. In other words, the team cannot offer a fixed target, but only a confidence level (the probability of achieving $115 million in savings is 90 percent). Confidence levels offer a better understanding of the overall project riskiness than simplified project buffers that are used in other companies.

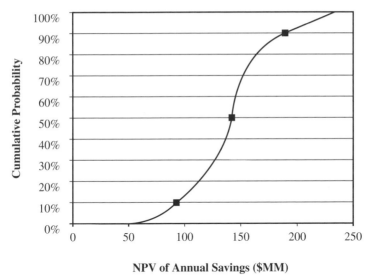

NPV of Annual Savings ($MM)

Figure 1.4 The PCNet project outcome distribution

Second, the value distribution curve offers a different dimension of priority for the key value drivers: The value distribution is influenced by the *variance* of the risk factors, and the distribution represents a "model" of how that variance impacts the project's value. This method is finer-grained than the expected impact that we have discussed above: Uncertain events often do not simply "occur or not occur" but have a continuum of possible values. For example, for the desktop migration subproject, the dominant risk lay in the prices set by the PC vendors. These prices had possible ranges, and the variance of the ranges defined the importance of the risk, both on the upside and the downside.

1.3.2 Risk Prioritization

Based on the impact assessment, risks are usually prioritized in order to allow the project team to focus attention on the most important ones. Typically, this prioritization is based on an "expected impact" of the risk, that is, size of impact, if possible in monetary terms, times probability of occurrence. However, one should not rely on this expected impact metric alone—a risk may have a moderate expected impact because it has a low probability; but at the same time, it may represent serious damage (a "showstopper") if it occurs. It may be advisable to pay keen attention to preventive or mitigating actions against such a risk.

In the PCNet project, hundreds of distinct risks were identified and listed in the risk planning process. They had to be prioritized in order to maintain focus (which was one of the main identified risks itself!) and efficiency. Rather than classifying the hundreds of risks themselves, the merger team chose to classify the subprojects in a type of ABC analysis according to their value (that is, the amount at stake when a risk occurred) and the probability of failure (aggregated from individual risks affecting that project). The logic of this analysis is shown in Figure 1.5. The project management team aggressively intervened in high-priority projects (high value and high risk), whereas projects with high risk but low value were delegated to the Operating Companies, with an offer to help if requested. The well-running projects were either watched (high value) or let run (monitored only on a routine basis).

1.4 Risk Monitoring and Management

Risk identification and assessment form the basis on which the project can proactively influence the risks and monitor them, responding quickly if they occur. The Metal Resources Co. IT merger illustrates this.

1.4.1 Proactive Influencing of Risk Factors

The identification of the most important risk drivers by the value distribution curves (Figure 1.4) allowed the project team to manage them proactively. For the PC prices, the dominant risk for the desktop migration, a team of project managers visited the vendors and aggressively negotiated in order to lock in prices at a low level. Not only did this reduce the savings variance (guaranteed prices for all countries were obtained), but the centralized negotiation also achieved prices that were even lower than hoped for, creating additional savings. Thus, the pricing risk turned from a downside potential into a substantial upside opportunity, which was successfully utilized.

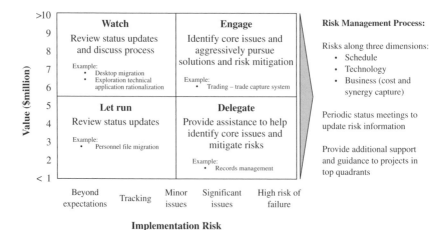

Figure 1.5   The Metal Resources Co. IT merger project prioritization

A second example relates to the considerable schedule risk of actually delivering all the desktops into all countries on time for the "new" system to start. Metal Resources Co. operated in countries such as Angola, Congo, or Armenia, where fighting might disrupt delivery, and where "gatekeepers," "consultants," or bureaucrats could block every move until permission is requested. Or sometimes they merely wanted to be shown attention and respect. In the aggregate, the schedule risk became large in such countries, or deployment might even be endangered entirely. So, for each country with significant bureaucratic restrictions, a plan was devised with countermeasures, emergency procedures, and appropriate buffers to allow for disruptions, and a deployment was not started until the remaining uncertainty had been reduced to a high confidence level such that the deployment could be carried out within the buffers allowed.

1.4.2 Monitoring and Reporting

Risk supervision concentrated on areas of high exposure. For example, the PCNet project came out as high priority from the classification in Figure 1.5: The desktop and server subprojects were in the upper right quadrant (high value and high risk), and the network project, while not directly of high value, was of high strategic importance and classified as high risk. The project team could not complain about a lack of attention from upper management.

In October 2002, work started in earnest, hitting multiple fronts at the same time: Telecom lines were rented and connected, network equipment was installed, security policies and software and directories were set up, and PCs were exchanged. Control of the large projects was paramount to keeping the integration on track and to producing the savings. An integration management office was set up for the merger as a whole (the "Integration Management Committee," of which Max Schmeling was a member), and the IT merger had its own integration office, the Project Management Office.

In monthly meetings, progress was tracked using the "Deployment Progress" monitoring tool (Figure 1.6). This reported on the progress status of PC deployment (for example, in January 2003, 1,428 of the 40,000 PCs had been migrated), sites with upgraded networks, reduced Internet access points to the global network, and reduced standard applications. The tool also showed the current budget status and offered comments on current events in the various regions of deployment.

In addition to the progress tool, which focused on operational figures, budgets and, of course, the financial synergies (or savings) were reported. The savings data were urgent: Only when they had been achieved and documented could they be incorporated into the accounting and bookkeeping systems, and then reported to external financial analysts. Being able to report booked synergies is very important for a CEO after an acquisition.

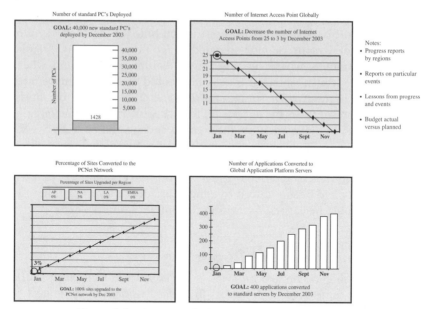

Figure 1.6   Aggregate deployment progress monitoring tool

During project execution, it became clear that the synergy progress reporting caused misunderstandings and tension. Real progress was made, not smoothly, but intermittently—for example, a site had to be completely migrated (up to three months' work) before savings really accrued, but then a large sum was saved at once. The synergy forecasts and targets, however, were "smoothed" and looked as if synergies were accruing at a regular rate every week. This caused an apparent deviation from the plan (a reporting artifact), as, for up to three months, it looked as if synergy capture was behind, before catching up. This required a lot of explanation: "Yes, although nothing has been booked, you have to trust us that the site migration is really on track and the savings will accrue as planned!" Reporting and education of the supervising committees had to go hand in hand.

1.5 Managing Residual Risks

The PCNet project had got off to a good start. However, in spite of all the thorough planning, "residual risk" began to plague the project. Residual risk took the form of nagging problems that kept arising out of the blue, none of them catastrophic in itself, but damaging nonetheless. In general, residual risk is not necessarily a sign of management problems: In a complex project, no planning, however thorough, can ever foresee all events; something not planned for will always happen. Therefore, it is key to building the capability of dealing with residual risk as it comes along. The PMI Standards Committee (1996) refers to this as "workarounds," and, if necessary, the PRM process is repeated if unexpected events occur.

In the PCNet project, the IT organization had managed on day one of the merger to build connectors between the two corporate networks, but there were no standards in place across the entire merged corporation. Thus, without rigorous change management processes in place, well-meaning people could (and did) introduce "tweaks" in, say, the e-mail system, unwittingly destabilizing an entire sector of the network. As a result, e-mail files were lost and messaging capabilities were corrupted. There were frequent small outages in some areas, which stubbornly persisted.

Moreover, some of Metal Resources Co.'s partner national companies suddenly demanded "local content," or "brokers," to be included in the channels of the hardware systems. These channel conflicts often caused delays and had to be mapped out and worked around, costing time and resources. A different problem occurred in Sri Lanka. The government partner, who was paying for the migration of 2,000 seats, decided that it would have to study the proposal thoroughly before giving its approval. This meant extra justification and a localized business case, again costing time and resources.

Unexpected problems came not only from the outside: Several business unit leaders within Metal Resources Co. slowed migration or postponed it from the original schedule, to avoid business disruptions or costs. One large European office threatened to delay a major deployment that had scheduling impacts on several other subprojects.

The Risk Management Office

To deal with residual risk, Max Schmeling built a formal structure, the Risk Management Office (RMO), complementing the Project Management Office (PMO). The PMO followed up on actions and reporting. The RMO focused on responding to deviations. It was a central control point, to which all teams were required to call in at least once a day to report on progress and problems that had arisen.

The RMO achieved two things. First, it represented a problem-solving resource—Metal Resources Co. had its own technical experts present in this center, plus experts on call from all technical areas at the main systems vendors (such as HP and IBM for PCs, Cisco for routers, Microsoft for operating systems, SAP for R3, EDS for the network operation, etc.), plus experts in culture and change management, who were also on call. Thus, when an unforeseen problem occurred, the center diagnosed it with the team in question and then helped to mobilize the expertise to bring about, or to plan, a solution as quickly as possible. Second, the quick information exchange offered a *fast spread of alarm bells (warnings) as well as of solution approaches*, across the many parallel teams. As many teams worked on very similar issues at multiple sites, a problem occurring at one site may well occur soon afterward at one or many of the other sites, and thus, warnings were relevant and a transfer of solutions was efficient. The quick communication of warnings from one team to another was dubbed "hot wire."

Thus, during each local deployment, a representative of the next local deployment team (in another state or another country) was present, so they became familiar with the logistical, as well as technical, issues. The Latin American deployment, for example, went very smoothly, thanks to this approach. Similarly, several problems that arose in the application migration to the new platform in Singapore were, once solved there, avoided throughout the Southeast Asian region.

Both the PMO and the RMO also attempted to prevent certain risks by enforcing strict standards (and thus reducing the complexity and number of things that could go wrong): For example, all of North America had to switch to a single SAP system configuration (there was a separate central control center for that project alone, which worked with all the organizational units to produce a common standard that satisfied most of the needs). Also, many technical and business software applications were standardized (such as statistical analysis packages, geological expert systems, etc.). Reducing the variety cut the number of different problems that could possibly arise and facilitated the spread of solutions across teams.

The RMO turned out to be very effective in responding to residual uncertainty. The problem of lost e-mails and corrupted e-mail capabilities, for instance, had to be attacked at two levels. The first level was technical—for example, when the lost files incidents were examined, the root problem turned out to be that Microsoft XP did not have a translator to automatically modify files. In response, the Microsoft developers made their own translator software available, which they used in-house. Installing this software systematically eliminated the problems and improved the overall network robustness. Several similar initiatives contributed to an overall network stabilization. The second solution level concerned change management processes: Over time, the merger team put such processes in place ("who can change what system features after discussing it with whom"), and over time convinced employees to comply with them, an initiative that eliminated incompatibilities introduced by local changes.

The Sri Lankan government partner eventually came on board, albeit at its own pace. This contributed to a six-month delay, but it did not "stop the show." The refining plant manager, who had refused the deployment, was won over with a combination of carrot and stick. The IT organization conducted a security audit at his site, which brought to light serious vulnerability to external attacks and other breakdowns. This allowed the team first to show him how awkward this might get for him locally (carrot) and, second, to make it clear to him that he would not be permitted to pose a risk for the rest of the organization (stick).

It turned out that the successful management of residual risks was crucial for the success of the project. The RMO was not simply a trivial variation on the general "risk monitoring and management" part of PRM. Installing the problem-solving capacity to respond to residual risks is fundamentally different from triggering ready-made responses to identified risks. Therefore, we think that the standard PRM methodology that is summarized in Figure I.1 must be enhanced by the explicit managing of residual risk. We present this enhancement in Figure 1.7: Residual risk management is a distinct activity that is applied in parallel to the risk management phase. In contrast to risk management, it responds to unexpected events that require unplanned real-time problem solving.

Our concept of residual risk management is consistent with Schoemaker's "strategic compass for profiting from uncertainty."[2] Schoemaker proposes a three-step process that prepares an organization for uncertainty. Although this approach is developed in a context of strategic planning, its philosophy has commonalities with our idea of handling residual risk: First, develop multiple future scenarios (to gain insight about the possibilities and uncover opportunities), develop a flexible strategy (that uses real options, analogous to the decision trees' ability to picture flexibility), and then monitor in real time and adjust to the unfolding future in a timely fashion (modifying the plan as you go along).

1.6 Learning and Sharing Across Projects
1.6.1 Learning over Time

The activities of the RMO enabled the organization to work with budget and schedule variances (deviations) in a more sophisticated way. For example, they performed variance analysis. There was significant overspending in Phase 3, because some work originally planned for Phase 4 had already been carried out at this point, and also because of many small "design changes," or improvements in protocols and processes that the organization implemented during the project. The activities of the RMO provided explanations for and documentation on residual risk and the respective responses.

Thus, the organization had a trace that allowed a thorough explanation of deviations, and an institutionalized effort to learn from the changes. One example of learning is the following: The early PC deployments took several man-days as the migration team was learning and stabilizing the components of the network. The later deployments required only a few hours (a reduction of 75 percent) and were much more stable.

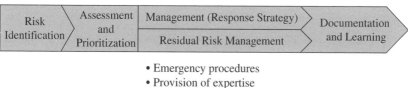

- Emergency procedures
- Provision of expertise
- Real-time problem solving
- Adjustment of procedures
- Communication

Figure 1.7   PRM process enhanced by residual risk management

1.6.2 Learning from the Residual Risks

The solutions that the RMO developed were systematically generalized and transferred to other problems, which meant that the organization developed a codified and explicit set of solution methods. In other words, the organization learned from experience. As an example, the change management processes introduced in the context of the e-mail system stability enabled the IT organization to better manage the system.

Similarly, the cajoling and convincing of the refining plant manager was then crystallized into a standardized, compelling argument used in all interactions with operating managers who thought they had no time for offline activities like IT migration (Figure 1.8). The argument again combined the carrot and the stick: On the one hand, it explained the benefits to the operating units themselves and emphasized the fact that they could get help. On the other hand, the document threatened stakeholders with the withdrawal of support for their network if they did not migrate. This standard argument was, of course, accompanied by personal visits and face-to-face explanations.

1.6.3 Overall Success of the PCNet Project and Learning Applied to Other Projects

The IT organization learned how to execute the merger without disrupting ongoing operations, how to apply state-of-the-art methods, and how to deal with residual risks. In the end, no unexpected event was serious enough to break the project. The thorough planning, combined with the flexibility of the RMO and the hot wire, was so powerful that the huge IT merger undertaking became a convincing success.

The total IT merger project beat its target by $20 million, producing $230 million of synergies in the first year, and the PCNet project made a significant contribution to this overfulfillment (partially driven by an extra $10 million in PC discounts, the risk upside that came out of the proactive negotiations). Overall, the project remained slightly under budget, although it took six months longer than originally planned.

Documentation, in the form of a template, was produced, listing the major risks (downside and upside) that occurred. This template was applicable to other IT merger projects and continued to be used in the company.

1.7 PRM as a Method and as a Mind-Set

This chapter illustrates, we believe, that PRM is a powerful set of methods that help project organizations to anticipate and respond to risks. However, PRM is not something that an organization can simply decide to adopt overnight. As an illustration of this, we have worked with a number of organizations that could not take full advantage of PRM because management did not have the necessary mind-set.

The PCNet Deployment Consultant team presents . . .

The Top Ten List of *"Reasons why you should quickly and carefully decommission your legacy IT environment."*

10. Dual environments will make it more difficult to maintain IP compliance, particularly once Microsoft ceases support of NT 4.0.
9. Dual environments are impacting our networks due to unnecessary traffic from the legacy infrastructures such as file replication, Exchange Global Catalog replication, SMS inventory and package traffic, as well as WINS and DNS traffic.
8. Increased vulnerability to security attacks and viruses as vendors start dropping maintenance support for WIN9x, NT4 and W2K, and our internal, centralized efforts are no longer funded for these environments.
7. Increased cost for support as troubleshooting by support staff becomes a lot more complex due to having to follow separate processes and using different tools in order to support two environments. Cost also increases due to reduced reliability and increased break/fix calls as hardware has lived long past its planned life cycle.
6. Legacy Master Account X1 and X2 domains will be decommissioned, leaving unreliable domains. The old PC and workstation environments will lose connectivity. There will also be performance issues as Master Account domain controllers are removed.
5. The decommissioning effort is part of Metal Resources Co. and RBD synergy cost-savings and the realization of these savings now becomes our responsibility.
4. The business case for the synergies will be compromised by having to support dual infrastructures.
3. Manpower can be redirected toward strategic projects once the deployment and decommissioning efforts are completed (and we can take our vacations now!).
2. Old computing standards monthly cost will be increased by x2, x4, x6 the longer you keep your old hardware. Costs to maintain old infrastructure will be divided by the number of remaining old standard users.

And the #1 Reason is...

1. The old desktop **has** become "non-standard." Yes, it is true. The sun has set on the old standard, with the IT design team only providing Norton Anti-Virus updates and major security patches. *Having old standard machines at your site makes your site "Non-Standard."*

   Here are three documents to help you in your efforts to decommission:
   Decommission Legacy Systems Guide
   Decommissioning Server Assets
   Decommissioning Workstation Assets

If the thought of pulling the plug on your favorite Compaq Proliant server is giving you nightmares and sleepless nights, then please e-mail me back about getting the PCNet Deployment Consultant team to offer decommissioning consulting services at your site.

Figure 1.8   Communication document for operating company compliance

Figure 1.9 (top) illustrates the mind-set with a metaphor: The traditional critical path mentality assumes that there is a well-defined target, and a path to get there, and the project must reach it; if obstacles are encountered or hostile winds distract the team, they'd better work harder at reaching the target or they will not have performed well. Take, as an example, the statement by one high-level manager who announced to his organization: "We need people who are reliable and fulfill targets. There are too many excuses

The Critical Path

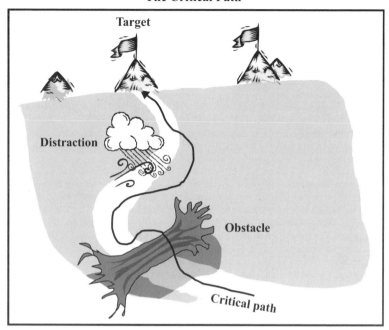

The Spirit of Risk Management

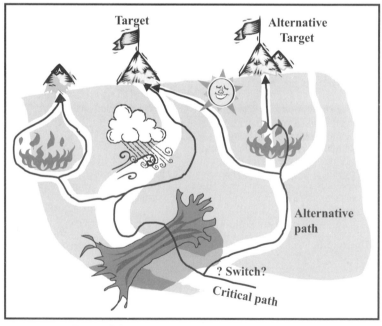

Figure 1.9   The critical path versus PRM mind-sets

and complications around here; we need people who simply get things done." In such organizations, allowing contingent actions and flexible targets makes management feel as if they are losing control. Project managers are "shot down in flames" when they propose contingent actions.

The bottom of Figure 1.9 illustrates the changed mind-set. Obstacles, alternative paths, and even alternative targets are identified and outlined at the outset, and a switch to the preventive/mitigating or contingent path (action) is triggered when the monitoring indicates that an obstacle has indeed occurred. Thus, PRM requires higher management sophistication— with PRM, targets and "getting things done" are fuzzier than a traditional organization is used to. Plans are more complex, monitoring is more difficult and subtle, and management must be sufficiently knowledgeable to understand when and why contingent action has been triggered.

Even more difficult is instilling awareness of residual risk (for example, the hostile wind that blows the project off course in Figure 1.9). A management that has not acquired the PRM mind-set, and does not understand the presence of residual risks, may revert to the critical path mind-set, punishing the team for running into trouble. We have seen organizations where the project teams did not trust supervising management to stay the course and built gigantic buffers into their plans, as a "private insurance" against residual risks. Sometimes, a game of "I cut your buffers" (management) versus "I build even bigger buffers" (project team) ensues, resulting in a total loss of planning and control.

The PCNet example in this chapter illustrates the power of PRM when it is competently executed, and when management uses it constructively. In the pharmaceutical, engineering services, chemical facilities, and power generation industries, and a few others, PRM has developed into a powerful way of guiding risky projects to success.

1.8 Summary and Conclusion

This chapter sets the baseline for this book by surveying the state-of-the-art in PRM. We have seen that PRM is a powerful method of achieving the stated project goals despite the risks. The method consists of four conceptual steps: (1) risk identification; (2) risk assessment and prioritization; (3) risk management with a collection of preventive, mitigating, and contingent actions; and (4) knowledge accumulation and transfer.

In the example of the Metal Resources Co. PCNet project, we saw the power of PRM in practice. In particular, the following lessons stand out:

▲ Thorough planning and anticipation is the foundation on which successful project management rests.

▲ Max Schmeling established responsibility for clearly defined outcomes, broken down to the task level. This accountability is the basis for project leadership.

▲ Risk planning is an integral part of project planning. Preparedness for deviations is the key to the team's ability to stay the course.

▲ Risk prioritization allows the team to stay focused. Focus is critical—once it is lost, the team will get bogged down and the project will stall.

▲ Transparency and clear communication of risks and progress status allows the team to remain coordinated and maintain a common direction. In a complex project, it is difficult to communicate risks, prioritization, and status in an easily understandable manner, but it is worth investing time and effort to accomplish this.

▲ Triggering planned responses to anticipated risks is not enough in challenging and complex projects. Residual risks will inevitably arise, as it is impossible to anticipate everything. Managing residual risk requires investing resources in real-time problem-solving capabilities (the RMO in the PCNet project). The standard PRM process must be enhanced by residual risk management (Figure 1.7).

▲ Successful PRM and residual risk management is not simply a method that can be routinely applied. It requires a management mind-set, in particular, the willingness to deal with unforeseen events constructively, without giving up the overall direction, and without blaming people.

▲ Learning and systematizing project experiences can improve execution, even within the same project, and can certainly benefit future projects in the same area.

We conclude with the question of whether the powerful method of PRM is the tool that can handle all projects, no matter how novel and complex. The answer, unfortunately, is no. The fundamental philosophy of PRM is still the achievement of stated goals, with a roughly agreed-upon approach, although it may change in detail. However, projects with high novelty (for example, a new market or a new technology) or high complexity (for example, many players with different sets of expertise, technologies, and expectations) may not have a (even roughly) defined target or approach. The rest of the book focuses on this added challenge. Chapter 2 illustrates an example where PRM is not enough. Chapter 3 begins to extend the PRM toolbox and mind-set.

Endnotes

1. This chapter is based on Loch 2005.

2. Schoemaker has summarized his approach in his book *Profiting from Uncertainty*.

The Limits of Established PRM: The Circored Project[1]

2.1 Early Design of the Circored Technology
2.1.1 Cliffs' Strategic Business Idea

Cleveland Cliffs, based in Cleveland, Ohio, is one of the largest iron ore and iron ore pellet suppliers to blast furnace integrated steelmakers in the United States. In the late 1980s, Cliffs' management observed a demand shift from blast-furnace-based steelmaking to electric arc furnaces (the newly fashionable "mini-mills," such as Nucor). But they had no product for the mini-mills, which used mostly scrap as their iron source.

Scrap contains many contaminants, or other metals included in the products from which the scrap is made. Therefore, the mini-mills needed clean iron inputs, the so-called direct reduced iron (DRI). This would allow producers, such as Nucor, to go head-on profitably against the integrated mills, even for high-quality steel. Thus, the demand for DRI was expected to rise steeply, from 17.7 million tons in 1990 to double the volume by 1997. Almost half of the DRI came from Latin America, with its cheap natural gas and huge iron ore reserves; three million tons came from Venezuela alone.

At the end of 1992, Cliffs' management decided to develop their own DRI business. They approached several potential partners—three steel companies and one iron ore producer—and entered discussions with them about sharing the risks and the rewards of an attempt to move into the DRI business.

Cliffs set up a task force, led by Senior Vice Presidents Bill Calfee and Dick Shultz, to analyze the available options. The task force examined different kinds of process technologies for converting iron ore into DRI. The Midrex technology was the "classic" and dominant process of producing DRI, accounting for 61 percent of world production in 1990. As Cliffs felt they were a latecomer in DRI and needed to differentiate themselves, they decided to pursue a new technology.

Two new technologies looked interesting but could not be pursued further because the technology partner worked with someone else or because of licensing problems. Thus, by late 1994, the task force revived earlier discussions with the German company Lurgi Metallurgie GmbH. In April 1995, John Bonestell, head of Lurgi's U.S. ferrous metal unit, met Bill Calfee at an iron ore conference in Vienna and explained the new Circored process to him. Bonestell arranged for Calfee to meet Walter Schlebusch, then Managing Director of Lurgi Metallurgie. They met on May 2, when Schlebusch said, "We are interested. How would you foresee an agreement?" Calfee drafted three pages of what he thought could be a development agreement, prompting Schlebusch to react: "We could live with this!" The exchange led directly to a series of phased tests, cofinanced by the partners, and ultimately to the go-ahead.

2.1.2 The Circored Technology and Early Preparations

Lurgi Metallurgie GmbH, a subsidiary of the MG Technologies AG, was a metallurgical process engineering company with a long history and widely acknowledged expertise of working with circulating fluidized bed (CFB) processes. Since the 1950s, the company had fluidized solids by blowing high-pressure gas into them and circulating the fluidlike mass through a reactor and a cyclone. The circulation offered high mixing and, thus, a fast chemical reaction. The company successfully applied the principle in material processing, waste incineration, ore processing, and energy production.

In the late 1970s, Lurgi had developed the coal-based Elred process. The gas-based successor process to Elred was given the name "Circored." In a climate of increasing interest in DRI, in 1994, Lurgi developed a process proposal and a lab pilot under the leadership of Dr. Martin Hirsch, the "brain father" of the CFB principle. The Circored process was simple and elegant, using only hydrogen to reduce the ore, and briquetting the DRI to produce hot briquetted iron (HBI). Lurgi argued that Circored was more efficient than Voest Alpine's FINMET (the most-talked-about new technology), achieving, with two reactors, the same output as FINMET with four. Thus, physical plant size and capital requirements would be lower for the same capacity, maximum plant capacity would be higher, and, working at lower temperatures, the process would be more robust. In contrast to Midrex, Circored used fine ore instead of iron pellets, saving an expensive pelleting facility amounting to $15 per ton. A simplified Circored process diagram is shown in Figure 2.1.

When Bill Calfee talked to John Bonestell again in May 1995, Lurgi could show first test results that demonstrated excellent quality and efficiency performance in a test reactor that processed 20 kg/hour continuously. Cliffs financed two further rounds of tests in the summer of 1995 (in return for a share in the royalties of subsequent Circored plants), at a total cost of about $1 million. The tests produced excellent results.

That summer, a Lurgi engineering team prepared the preliminary design of a proposed Circored plant (process calculations including dynamic process modeling, flow sheet development, layout, and cost estimates). Based on this effort, Lurgi made Cliffs an offer that was more detailed than usual at this early stage. The team had even more detailed piping and instrumentation diagrams available internally.

In the fall of 1995, Lurgi organized a conference with Dick Shultz and Bill Calfee from Cliffs and academics from several German engineering schools. The participants discussed the engineering challenges of the technology and concluded that there *were* risks, but that they were *known* and could be overcome. (The upper part of Figure 2.2 shows a simplified excerpt of the risk lists.) After this thorough risk assessment, they were convinced. While everyone was aware that the facility would be a first-of-its-kind, the technology looked straightforward, Cliffs had (in their mines) experience in the construction of large capital projects, and Lurgi had the process expertise. The project gained momentum.

Description of the Circored Process

The basic chemical reaction equation of the Circored process is $Fe_2O_3 + 3H_2 \rightarrow 2Fe + 3H_2O$. A design capacity of 100 tons/hour of iron ore fines are dried and preheated in the preheater (Figure 2.1, left-hand side) to a temperature of approximately 850° to 900°C: High-pressure air is blown into the solid at the bottom, and the highly mixed air-ore fluid circulates through the recycling cyclone and is efficiently heated. High pressure is also needed in the first seal pot to ensure fluidization. Through the second seal pot and the bucket elevator, the hot material is introduced into the four-bar pressurized hydrogen atmosphere in the middle lock hopper (the hot material is introduced in batches, using valves above and below the middle hopper to prevent pressure being lost).

The ore has now entered the reduction part of the process, a pressurized hydrogen atmosphere of 650°C (Figure 2.1, center). This temperature is low enough for the iron not to stick together in clumps or to the chamber walls, but high enough to ensure an efficient reaction. In the first-stage CFB reactor, the material is again fluidized with high-pressure hot hydrogen blown into the reactor and the seal pot, and circulates through the reactor and the seal pot. In this perfectly mixed state, 70 percent reduction is achieved very quickly, with a retention time of about 15 minutes. After prereduction, the material is discharged into the secondary fluidized bed (FB) reactor. Here, the material is also fluidized by hot hydrogen blown in from the bottom, but it does not circulate. The fluid flows through four chambers, bubbling over the top of and through holes in the walls. The longer the retention time, the higher the extent of reduction. After 3 to 4 hours, a metallization of 93 to −96 percent is achieved, which is an excellent quality for steel production.

If the steel plant was next door, this would be the end of the process. After depressurizing, the high-quality DRI, exiting in the same form as the ore input (fine sand), could be inserted into a steel furnace.

However, DRI is unstable and unsafe for transportation—pure iron reoxidizes even at room temperature and may cause fires on a ship. Therefore, the DRI must be pressed into hot briquetted iron (HBI), high-density iron bars that have such a small surface to the air that they do not reoxidize. Briquetting requires a temperature of at least 680°C to obtain high-quality HBI (with a density of greater than 5.0 g/cm). Thus, the DRI is discharged from the FB reactor into a flash heater. Preheated hydrogen is blown from the bottom into this vertical shaft at high velocity and carries the DRI to the top, heating it to about 700° to 715°C.

Now, pressure and hydrogen are removed in a reversal of the initial lock hoppers: Through three sealed hoppers, the material is dumped, hydrogen is replaced by nitrogen, and then the pressure is released (Figure 2.1, right-hand side top). Hydrogen is extremely explosive, and great care must be taken to ensure that none of it leaks out of the closed circuit in order to guarantee safe operation of the plant. Each of three briquetting machines (Figure 2.1, right-hand side bottom) contains two wheels that turn against each other, each wheel having the negative of one half of a briquette on its face. The DRI is poured onto the wheels from the top and is pressed into briquettes, or iron bars.

Energy for the endothermic reduction reactions is supplied by heating both the ore and recycle gases. The off-gas from the recycle cyclone of the reactors passes through a process gas–heat exchanger and a multiclone for the recovery of ultra-fine dust particles, which are recycled into the last compartment of the FB reactor. The off-gas is then scrubbed and quenched simultaneously for the removal of dust and water produced during reduction (the chemical reaction produces 30 tons of water per hour). The cooled and cleaned process gas is then recompressed and subsequently preheated in gas-fired heaters to a temperature of approximately 750 °C before being reintroduced into the reactor system.

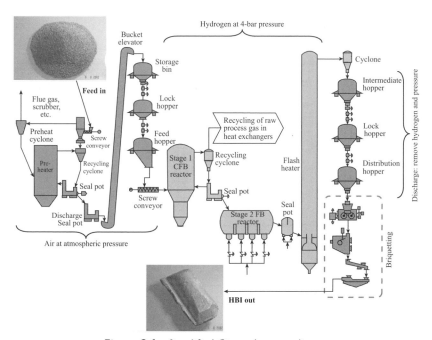

Figure 2.1   Simplified Circored process diagram

2.2 Joint Venture and Business Plan

As the project with Lurgi became more and more concrete, Cliffs pursued a joint venture structure for the facility. Of the originally identified partners, only LTV Steel, desirous of using DRI in a mini-mill they were building in Alabama, signed up. Shareholder negotiations for the Trinidad joint venture, called Cliffs and Associates Limited (CAL), went on through the second half of 1995. Cliffs wanted Lurgi to take an ownership position because Lurgi refused to give a performance guarantee for the plant—in general, process engineering companies are reluctant to give guarantees on first-of-its-kind technologies because the risks incurred would be too high for the usually small engineering suppliers who live on margins of about 3 to 5 percent. With part ownership, Cliffs hoped to give Lurgi an incentive to do their best to make the plant work. Lurgi reluctantly gave in. The shares in CAL were 46.5 percent each for CCI and LTV Steel, and 7 percent for Lurgi. Lurgi paid for its share with a cash injection of $6.8 million. A shareholder agreement was drafted in November 1995.

In parallel, Bill Calfee negotiated the ore supply from the Brazilian ore producer, CVRD, and finalized a preferred-price gas supply contract with the Trinidadian government. Also, an operating contract was drawn up that specified how CAL would be managed (Cliffs would get a management fee of 0.8 percent of revenues from CAL) and how the HBI would be sold (through another subsidiary of Cliffs).

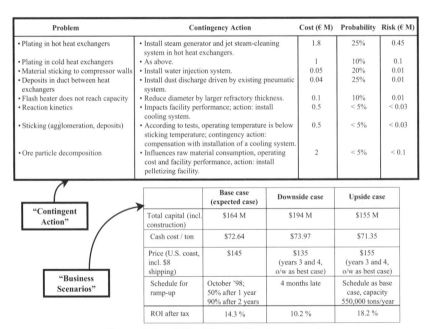

Problem	Contingency Action	Cost (€ M)	Probability	Risk (€ M)
• Plating in hot heat exchangers	• Install steam generator and jet steam-cleaning system in hot heat exchangers.	1.8	25%	0.45
• Plating in cold heat exchangers	• As above.	1	10%	0.1
• Material sticking to compressor walls	• Install water injection system.	0.05	20%	0.01
• Deposits in duct between heat exchangers	• Install dust discharge driven by existing pneumatic system.	0.04	25%	0.01
• Flash heater does not reach capacity	• Reduce diameter by larger refractory thickness.	0.1	10%	0.01
• Reaction kinetics	• Impacts facility performance; action: install cooling system.	0.5	< 5%	< 0.03
• Sticking (agglomeration, deposits)	• According to tests, operating temperature is below sticking temperature; contingency action: compensation with installation of a cooling system.	0.5	< 5%	< 0.03
• Ore particle decomposition	• Influences raw material consumption, operating cost and facility performance, action: install pelletizing facility.	2	< 5%	< 0.1

"Contingent Action"

"Business Scenarios"

	Base case (expected case)	Downside case	Upside case
Total capital (incl. construction)	$164 M	$194 M	$155 M
Cash cost / ton	$72.64	$73.97	$71.35
Price (U.S. coast, incl. $8 shipping)	$145	$135 (years 3 and 4, o/w as best case)	$155 (years 3 and 4, o/w as best case)
Schedule for ramp-up	October '98; 50% after 1 year 90% after 2 years	4 months late	Schedule as base case, capacity 550,000 tons/year
ROI after tax	14.3 %	10.2 %	18.2 %

Figure 2.2   CAL risk lists and business scenarios

On March 12, 1996, Calfee and Shultz presented to Cliffs' board of directors a proposal to build a 500,000-tons/year plant in Point Lisas, Trinidad (Figure 2.3). This represented a scale-up of a factor of 5,000 compared to the tests, from a 20 kg/hour lab reactor to a 100 tons/hour plant. The maximum feasible capacity for the Circored process was 2 million tons/year. Thus, the proposed plant was sized in between a pilot facility (which might have had a 50,000-tons/year capacity) and a scale-efficient plant. This represented a compromise between a pure pilot that would be far from paying for itself (for which both Cliffs and Lurgi had decided they did not have the money) and the excessive risk of a huge first-of-its-kind facility.

The business case is shown in the lower part of Figure 2.2, with an upside and downside scenario, a further application of PRM. With prices at $155/ton over the previous two years, the consensus forecast by all external analysts was $145/ton delivered to the U.S. Gulf Coast with a range of plus-or-minus $10/ton. They felt that there was plenty of demand, as most of the growth in steel production was concentrated in mini-mills and scrap was scarce—Cliffs would sell all 500,000 tons; the only stumbling block was the price. The expected after-tax ROI of 14.3 percent was not sensational but was respectable for a small-scale plant, and the parties agreed that real money would be made when a follow-on facility of 2 million tons/year capacity was built next door, immediately after ramping up the first facility.

The boards of Cliffs, LTV Steel, and Lurgi approved the high-level technical plan, based on the thorough preliminary engineering, and a capital expenditure of $170 million.

Figure 2.3 The CAL Circored facility in Point Lisas, Trinidad

2.3 The Construction Phase, May 1996–April 1999

Cliffs put a wholly owned company in place to supervise the construction work. In parallel, Cliffs Associates Limited (CAL) was set up as the later operator of the facility, with General Manager Ray von Bitter. He was the general manager of Cliffs' big Minnesota ore mine, with sound experience in installing and managing capital-intensive equipment.

Cliffs selected Bechtel Canada as the EPCM (engineering, procurement, and construction management) contractor who would supervise the construction companies. They had a large presence in Trinidad, understood Trinidadian labor, and had experience working with Cliffs. Because of the newness of the technology, Bechtel insisted on a cost-plus-fee basis.

Lurgi was to supply the entire core of the plant as a subcontractor (all except loading and unloading, and heat and process gas supplies—this was their proprietary technology). Lurgi even asked for the contract for the whole plant, but while Cliffs trusted Lurgi's technology, they did not know how good Lurgi was as a construction manager. Lurgi was awarded a lump-sum contract of $56 million for the core equipment, with a contingency of $3.1 million (in addition, a separate contract worth over $12 million was awarded a year later to Lurgi Oel Gas Chemie GmbH, who was the lowest bidder for the hydrogen reformer plant). Both contracts being fixed price, Lurgi carried the risk of cost overruns up to 10 percent of the contract value. CAL would commission and start up the plant, with technical assistance from Lurgi.

The groundbreaking occurred in February 1997 in Point Lisas, in a festive ceremony. The project schedule was very tight for a first-of-its-kind plant: Mechanical completion (end of construction, erection, and cold/function testing) was planned for October 1998, 27 months after the start of process engineering, and operations (end of hot commissioning) were to begin in December 1998. After a one-month grace period, delay penalties would set in for Lurgi.

CRIMCO would have preferred to deal with only one contractor, Bechtel, viewing Lurgi as an equipment supplier. But Lurgi did not accept that approach—they saw the core plant as *their* baby. Tensions between Lurgi and Bechtel started immediately and lasted throughout the project. Lurgi had reservations about cost-plus-fee contractors, and the two parties competed and wrangled over who was given the right to perform engineering on what project part (for example, the weight-bearing steel construction of the core plant). In addition, Bechtel, as the general contractor, exerted cost pressure on Lurgi.

A number of project management problems dogged the construction phase. Bechtel underestimated the total erection cost; all the bids came in twice as high as estimated. Bechtel, in its role as general contractor, was then allowed to award the lump-sum construction contract to themselves (a different office), in order to maintain the cost estimates. This contributed to a suspicion that Bechtel was protecting itself against overruns by aggressively logging small changes in the project to claim additional fees. Indeed, the year 1998 brought about a $20 million lawsuit in which Bechtel claimed additional fees.

Bechtel and Lurgi continued to criticize each other as errors occurred on both sides. For example, Bechtel underestimated the difficulty of transporting the huge 400-ton FB reactor to the site and had to build a sea landing specifically to unload it. Lurgi experienced its own problems with the core plant steel structure, which it had outsourced to German suppliers. Coordination failures among these subcontractors resulted in some steel structure components not fitting, and expensive and time-consuming rework having to be performed on the construction site.

The collaboration was not made easier by the fact that Bechtel and Lurgi had different problem-solving styles: Bechtel emphasized procedures and sign-offs, reflecting their tradition of large, cost-controlled projects, while Lurgi emphasized decentralization and local problem solving, reflecting their tradition of building facilities under unique and new circumstances. This inevitably caused daily clashes. Moreover, people from three nationalities had to work together on the site—Americans, Germans, and Trinidadians. Their different communication styles and work habits caused further tensions.

Cliffs also contributed its share of problems. Its overseeing procedure was too weak, as the responsible project manager remained in Cleveland, visiting the site only every few weeks. This had been enough in normal mining projects, but now it prevented the project manager from reacting fast enough, given the turmoil of the project. In the end, all these problems caused "mechanical completion" of construction to be delayed by six months, from October 1998 to April 1999 (see the project timeline in Figure 2.4).

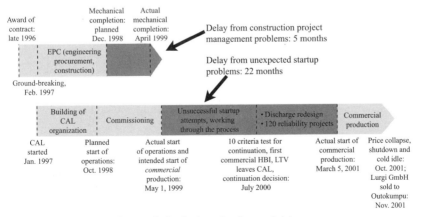

Figure 2.4 Project timeline and delays

2.4 The Startup Phase, May 1999– Summer 2000

Finally, mechanical completion was declared in April 1999, and Bechtel was released. On May 1, CAL took over responsibility from CRIMCO, and the startup team (run by Lurgi) heated up the facility with hydrogen. On May 9, they charged ore to test the whole facility, at a lower pressure of 2 bar and a lower temperature of 500°C. This produced great news and excitement. After only one day, they successfully produced DRI, which they dumped on a storage pile. On May 13, an operator erroneously poured DRI into the briquetting machine, and they even produced a small amount of HBI (although of low quality because of the insufficient temperature, and endangering the briquetter by putting the wheels under stress: DRI is less formable at the lower temperature). That night, champagne flowed to celebrate the success. The rest should follow quickly now.

The celebration was short-lived. From then on, bad news came in quick succession. Four days later, hot spots appeared on the reactor shell (because hot gas had migrated through the refractory), and the refractory in the FB reactor collapsed.[2] This shut the plant down for six weeks. Then, a shaft was bent in a fan for the preheater. The screw feeder going into the reactor leaked hydrogen through the seals causing a major fire, and the plant had to be shut down for three weeks to fix it. The main header of the gas heater, made of an exotic material, cracked and could not be welded— causing the plant to be shut down again. Then, the water depressurizer failed: 30 tons/hour of water were a by-product of the process and had to be relaxed to atmospheric pressure. Above 2 bar, hot hydrogen "shot through," and dust at the bottom seal pot regularly blocked the depressurizer.

And so it went on. Ultimately, a major part of all equipment experienced problems and had to be repaired or replaced, in many cases not because of bad workmanship, but lack of knowledge about the design requirements placed on the equipment by the hot, aggressive, and abrasive material. As the team tried to bring the plant up to design temperature, they found that the texture of the material changed from that of sand to that of molasses, more and more viscous and sticky. The material would coat the internal surfaces of all the valves and "choke" them. Eventually, all the valves had to be redesigned, and in the intervening months, the existing valves had to be "pampered" and cleaned frequently.

Finally, once the team worked their way through the process to the end, the discharge system into the briquetting machine was quite simply found not to be working. The hot DRI at 700°C just did not flow; it got stuck in the discharge hoppers virtually every hour, and no continuous flow was possible. Moreover, the little HBI that *was* made was "fish-mouthed," meaning that parts of the briquettes stuck on the briquetting wheels, giving the briquettes a less-than-perfect look because of holes in their sides and endangering the machines.

The problems went on and on, surfacing one by one. As the plant contained hot hydrogen under pressure, it was, essentially, a bomb. Every problem forced the team to replace the hot hydrogen with inert nitrogen, cool the process down, do the repairs, and then heat it up again, only to run into the next problem. What exhausted everybody was constantly hearing, "We have solved this problem; now we are almost there, and we are going to make it," only to encounter another problem the next day. With the never-ending problems, the team felt that they had lost credibility, and dedication was burned out. Some felt that the pressure on individuals mounted too high, with personal costs on all sides.

2.5 A Management and Design Change

At the end of 1999, Ed Dowling, senior vice president of operations at Cliffs, was asked to go to Trinidad and fix the problems. He and Christof von Branconi, the successor of Schlebusch as managing director of Lurgi, put together a task force of experts from Lurgi, Cliffs, and external experts to investigate the last, and most intractable, problem of flow stoppages in the discharge system to briquetting.

In April 2000, Dowling appointed a new general manager for CAL, Steve Elmquist. Ray von Bitter became head of the discharge system task force, with the mission to come up with a design by the summer. Elmquist performed an audit on every piece of equipment, creating a punch list of the most important problems. This resulted in 120 reliability improvement projects, which essentially summarized the remaining problems after the efforts of the previous year.

In parallel, Lurgi produced a new design of the discharge system: a vertical pipe, with a seal pot at the bottom into which hydrogen was blown for fluidization and to drive out the hydrogen, and then a rising pipe within which the pressure was relaxed. This design was successfully piloted at the labs of two external engineering companies.

By June 2000, nerves were frayed within both Cliffs and Lurgi. They had still not produced sustained runs. Then another blow came: LTV decided that they had had enough and wanted to exit from CAL. LTV had suffered severe financial troubles since 1999 and were close to declaring bankruptcy. They proposed selling their 46.5 percent share to the remaining partners for $2 million (plus additional payments when DRI prices would recover).

In June 2000, CAL commissioned a report on the viability of the plant by a consulting company. The report argued that the market outlook was marginal, but more seriously, it saw some grave technical risks. In particular, residual hydrogen traces in the briquettes might make the product dangerous. The report concluded that the plant should be shut down. But Dowling and Branconi disagreed with the conclusions. They tested the briquettes and found no hydrogen traces. Still, there was little confidence left at Cliffs and Lurgi that the plant could ever run. So, Dowling commissioned a three-week test run in July (at a cost of $1 million), without the new discharge system in place. The run was to prove the fundamental viability of the process. Ten goals were formulated that had to be met in order for the plant to get a chance to continue. While the discharge hopper valves were still jammed up because of sticking DRI (implying that the new discharge system was badly needed), the test run did, indeed, meet all 10 criteria. Still, the facility was within a hair's breadth of being shut down.

Armed with the test results, Dowling and Branconi presented a new investment proposal to the combined boards of Cliffs and Lurgi in August: An additional investment of $45 million would allow the company to execute the new discharge system and to implement the 120 reliability improvement projects. The target startup date would be February 28, 2001. The data presented were convincing enough for the boards to accept the proposal. In addition, Cliffs and Lurgi were now the sole owners of CAL, with stakes of 82 percent and 18 percent, respectively.

By using stringent project management methods, the discharge system was completed on time. On March 8, 2001, at 4:38 P.M. local time, the first commercial HBI hit the ground under the briquetter. It was so exciting after such a long period of frustration that some employees just stood there and giggled uncontrollably. The following months saw a steady improvement of plant operations. During the three months of August through October, the plant managed several extended runs at the design availability of 85 percent for the whole system. This represented the system reliability that fulfilled Lurgi's obligations as a contractor. In the end, the facility had become a technical success.

2.6 Market Turmoil

In October 2001, DRI prices fell to their lowest point in recorded history, $75/ton, as the Manhattan World Trade Center attacks had pushed the world economy over the brink into a recession. At these prices, no one in the industry could make money. After a scheduled maintenance shutdown in October, Cliffs decided not to start the plant back up. Rather than permanently closing the plant, they "cold idled" it, implying a long-term shutdown while retaining the key personnel (about 90 people). The cost of the cold-idled plant amounted to $10 million per year.

In November 2001, MG Technologies sold Lurgi Metallurgie to the Finnish company Outokumpu. This led to von Branconi's departure. The negotiations between Cliffs and the new partner proved difficult, as Outokumpu was reluctant to share the cold-idling costs.

In the spring of 2002, DRI prices recovered somewhat, to $90/ton, thanks to changes in the world market supplies and to steel import tariffs imposed by the U.S. government. Still, in August 2002, the financial pain proved too great for Cleveland Cliffs to bear. The company decided to write off the Point Lisas facility, suffering a one-time restructuring charge of $108 million (a significant amount for a company with annual revenues of $450 million).

The end of the Circored facility seemed once again imminent, with the same fate looming that Nucor's competitive iron-carbide venture across the street in Point Lisas suffered—this facility had technically failed and was dismantled during 2002. Leo Kipfstuhl, CAL's CFO, attempted a management buyout during 2003, but without success.

In August 2004, the big break came. Cliffs reached an agreement with the International Steel Group (ISG) for $8 million plus assumption of liabilities and up to $10 million in future payments, contingent upon production and shipments. The license to the technology was transferred to ISG. ISG was a new steel company founded by private investors who had bought the defunct assets of LTV Steel and then bought bankrupt Bethlehem Steel at fire-sale prices. With assets bought at market prices and sweeping reforms in labor relations in force, ISG was able to operate profitably. Not only did they buy the Circored facility, but they also started planning for the expansion to the 2-million-tons-per-year target that had originally been foreseen as the fully efficient scale.[3]

Thus, the Circored plant seems, ultimately, to have achieved success and may yet have a significant influence on how the industry converts iron ore into pure iron, albeit five years later than originally planned, and under a different owner. As is often the case with breakthrough technologies, the original owner lost out financially, and got out. In addition, several competent people had their careers damaged by the events. The project manager and the board were surprised when the problem-solving activities turned out different from what had been predicted, and surprise turned into frustration and disappointment. The question is what was missing and what we can learn from this project that is applicable to PRM more generally.

2.7 The Limit of PRM: Unforeseeable Uncertainty

In diagnosing what had happened in the Circored project, we must start with two observations. First, the project management errors during the construction phase were avoidable, but they added, in the end, only four months to the schedule overrun. The majority of the delays, 22 months up to the successful ramp-up in March 2001 (see Figure 2.4), were attributable to the difficulty of getting the process to run.

Second, Cliffs and Lurgi had performed competent project risk management. The 1995 conference used combined experts to diagnose risks (Figure 2.2 summarizes some risk lists and business scenarios). The project plan included many contingencies, and even residual risks were tackled. Both Lurgi and CAL continued to refine their risk analyses all along, checking for new potential risk factors. Why was PRM not sufficient to handle the risks?

The short answer is that the plant represented a scale-up of a factor of 5,000 compared to the lab reactor that had been tested at Lurgi. In the words of Martin Hirsch, this meant that "the process kinetics were completely unknown. We simply did not know whether the circulating fluid would be stable. Nor did we know whether the stationary fluid bed would be sufficient for achieving 95 percent metallization. We were obsessed with retention time, which is why we made the walls in the stationary reactor too high. It turned out that we had plenty of retention time, but the high fluid level caused dust [that is, ore substance] to be sucked into the process gas circuit and cycling back to the CFB reactor. And so on. None of this could be calculated beforehand. I knew that because I had started up three novel process generations earlier in my career."

This implies that PRM simply was not able to predict the major potential problems. Problems had been identified, for sure, but most of these never materialized, while new ones appeared that were simply unforeseeable, and others, although foreseeable in principle, simply slipped through. As one Lurgi manager put it, "You face an entire forest, and each tree is a potential problem. In principle, you can focus on any one tree beforehand and study it, but you can't look at them all. You have to choose a few and focus on them, but then it may turn out that the relevant trees are totally different." In other words, it was not possible for PRM to proactively handle all uncertainty except for small residual risks, which could be handled with a risk management office commanding some extra resources. Rather, major problems were missed, not because of a lack of diligence but because of a fundamental lack of knowledge.

Thus, the CAL people, supported by the Lurgi team, were forced into a trial-and-error mode. They resolved one problem after another, slowly "working their way through the process," each time hoping it would now run, only to be disappointed. The Cliffs people, led by Ray von Bitter, found this extremely stressful. Although they knew it was a first-of-its-kind facility, their project experience fundamentally consisted of new equipment installations in mines. They were emotionally unprepared for the iterations that were required. Ray von Bitter suffered intensely when he had to announce to the board that the successful startup, thought to be around the corner at the last presentation, had again to be pushed back. This was, of course, not helped by the fact that only two months had been foreseen for commissioning and startup in the original plan.

The Lurgi team, in contrast, was also stressed, but they knew that the ramp-up would not be linear. Although this was the first major assignment for Peter Weber, the project manager, during the ramp-up phase, he was extremely well trained and had listened many times to the war stories of Hirsch, who had ramped up three first-of-its-kind technology generations in his 30-year career. But this was implicit, uncodified, and tacit knowledge that Lurgi had not known to communicate effectively, and Cliffs had not known enough to understand, in the initial negotiations. In addition, Weber was not in control—according to the contract Lurgi had with Cliffs, he acted only as an advisor, amassing profitable engineering hours for Lurgi. Responsibility and risk had passed to CAL.

The intuitive knowledge of the fundamental novelty of the facility made Weber and his team approach the design parameters very slowly—they went to 1-bar process pressure (at 500°C), then to 2 bar, in order to ensure that all hydrogen leakages were found and a potential explosion risk was avoided. The ballpark of the right combination of the 300 process control parameters was not known, so the team had to feel its way toward controls that would at least work. This caused additional tensions with the Cliffs engineers, who thought this unnecessary fiddling; according to the Cliffs experience, testing of a facility was best done by going to the design load straight away. This is, of course, true for a known process, or even the second plant of a technology: The ballpark parameters exist, and approaching the design load becomes much simpler.

Based on the misunderstanding of the nature of the uncertainty, the methodologies used by von Bitter and CAL were insufficient. They did not develop extensive documentation of the known state of the process and of conjectured problems. They tested the process steps linearly (because they always thought they were close), rather than running multiple experiments in parallel wherever possible.

Inadequate documentation prompted Ed Dowling to diagnose incompetence when he intervened in January 2000. The success of his 120 reliability improvement projects and the discharge system redesign confirmed for him that a lack of rigorous methods had been the cause of the problems. But while, indeed, methods had not been sufficiently developed, this assessment underestimated the true nature of the challenge. When he took over and successfully applied control and quality methods, he built on the work of the previous year, which had removed the fundamental unforeseeable uncertainty and had reduced the problems to a manageable number that were all visible and, in principle, understood. In other words, when Dowling stepped in, the nature of the project's uncertainty had changed from unforeseeable to structured, so PRM methods now worked.

But it was too late for Ray von Bitter, who went into early retirement in the spring of 2001. He commented: "The difficulty of the project was not clear when I was asked to do it. But even though you can't do anything about it, you're still punished. I guess that's called life in the big city."

2.8 Summary and Conclusion

We demonstrated the power of the PRM mind-set, and the methods associated with it, in Chapter 1. This PRM approach rests on a fundamental assumption, namely that we are operating essentially on *known terrain*, where it is known, in principle, what events and outcomes of actions to expect, and with *moderate complexity*, where the nature of the "solution space" is roughly known, where an action does not cause entirely unexpected effects in different parts of the project, and where we can choose a best course of action. In other words, we know the *range* of things that can happen and their causes, even if we may not be able to predict with certainty which of the identified events will happen, or have good probability estimates.

Not all projects fulfill this assumption, however. On the contrary, projects that are novel in terms of the technology employed and/or the markets pursued, and projects of long duration, are commonly plagued by fundamentally unforeseeable events and/or unknown interactions among various actions and project parts.[4] A "straight" application of PRM, without recognizing the additional novelty challenge, is insufficient and may even have destructive effects. In this chapter, we have demonstrated this insufficiency on a quite typical example of a novel project.

Endnotes

1. This chapter is based on Loch and Terwiesch 2002. For further background, see also von Bitter et al. 1999; S. A. Elmquist, E. C. Dowling, and L. A. Kipfstuhl. 2001; S. A. Elmquist, P. Weber, and H. Eichberger 2001.

2. Refractories are special masonry walls inside the reactor that protect the metal walls from the heat of the chemical process.

3. Events continued to evolve at a fast pace. In October 2004, a merger between ISG Steel and the Indian company, Mittal Steel, was announced, creating the largest steelmaker in the world. In November, regulatory approval was still pending.

4. For examples, see Morris and Hugh 1987, Schrader et al. 1993, Hamel and Prahalad 1994, Miller and Lessard 2000, Pich et al. 2002.

A Broader Look at Project Risk Management

3.1 Understanding the Fundamental Types of Uncertainty

We examined the power of the project risk management approach in Chapter 1, and we observed some limitations of PRM practices in Chapter 2. In order to understand the power of the PRM approach, and the roots of its failure, we need to understand and classify the *sources* of project risks, not only by the contextual source, as is done, for example, in risk lists, but also by the foreseeability of the underlying influence factors and by the complexity of the project. Figure 3.1 offers illustrations of the major concepts that we will be discussing in this chapter.

The top picture in Figure 3.1 illustrates the sources of risk that standard PRM methods are designed for: *foreseeable uncertainty* and *residual risk*. We saw in Chapter 1 how PRM embodies a mind-set of planned flexibility: Obstacles, alternative paths, and alternative targets are identified and outlined at the outset, and a switch to the preventive, mitigating, or contingent path (action) is triggered when monitoring indicates that an obstacle has indeed occurred. This may even include response to unforeseen, or "residual," risk via extra capacity (slack) or improvisational problem solving that allows the project to recover when small, unforeseen obstacles appear on the horizon.

Another source of project uncertainty is project *complexity* (illustrated in the middle picture of Figure 3.1). Project complexity can arise from either complexity in project tasks or stakeholder relationships. As we saw in the PCNet example in Chapter 1, the complexity of potential interactions in the e-mail system made it virtually impossible for the project team to predict the consequences of local changes to the system, resulting in lost e-mails and other system malfunctions. Thus, the project team had to implement rigid control over, and fast response to, local tweaks to the system in order to keep it within a known "control state."

The final, and most difficult, sources of project uncertainty are what the engineering community refers to as *unknown unknowns* (unk unks) (illustrated in the bottom picture of Figure 3.1). As we saw in the Circored project in Chapter 2, projects that are novel in terms of the technology employed and/or the markets pursued, and projects of long duration, are commonly plagued by fundamentally unforeseeable events and/or unknown interactions among different parts of the project.[1] A "straight" application of PRM, without recognizing the additional novelty challenge, is insufficient and may have destructive effects.

In this chapter, we classify sources of uncertainty and complexity in projects, discuss the current state of the art in PRM, and then preview what we propose should be done when a project must "manage the unknown."

3.2 Foreseeable Uncertainty and Residual Risk

We begin by observing that the standard PRM approach rests on a fundamental assumption—namely, that we are operating essentially on known terrain, where it is known, in principle, what events and outcomes of actions to expect and, with moderate complexity where the nature of the "solution space" is roughly known, where an action does not cause entirely unexpected effects in different parts of the project and where a best course of action can be chosen. In other words, we can *foresee* the range of things that can happen, and their causes, even if we may not be able to predict with certainty which of the identified events will happen or to what degree of probability they are likely to occur.

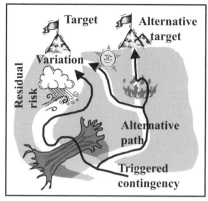

Variation, risk, and residual risk

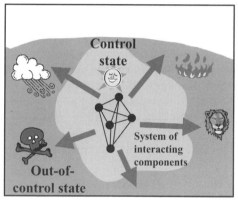

Maintaining a control state to avoid unk unks

Major unavoidable unk unks:
No known path but stakes in the
ground pretend we have one

Figure 3.1   The fundamental sources of project risk

We find it useful to consider two different types of foreseeable uncertainty, although theoretically, they are similar: variation and foreseeable events. We will then conclude this section with a discussion of residual risk, recognizing that there will always be some risk that escapes the initial risk planning process.

3.2.1 Variation

In many projects, it is not possible to identify and proactively influence all risk factors (partly because no historical data are available on which to base the estimates). The widening of the path in the top picture of Figure 3.1 depicts a situation that is virtually universal in projects: We can choose courses of action, but we do not know exactly what the values of the influence variables are and/or the nature of their impact on the final project status. Thus, the final project status can be planned only with "noise," represented in the picture by the widening of the path.

Variation in project performance makes the project outcome a nondeterministic event, a *range* of outcomes with probabilities. It is dangerous to pretend that this range of outcomes does not exist and to force teams to commit to deterministic targets. Forcing a deterministic answer to a stochastic problem often causes people to cover themselves and become overly conservative in their estimates. Project managers have long known this and have developed two methods for highlighting and managing variation: simulation techniques and project buffers.

Simulation and the Communication of Uncertainty
Consider the example of a traditional project plan, usually depicted in the form of a Gantt (or bar) chart, or equivalently, as a so-called network flow diagram or activity network. An example is shown in Figure 3.2, which depicts the plan of a 10-week project for preparing the bid for the development of an unmanned aerial vehicle (UAV), the type of small automated flying vehicle that was used for reconnaissance in the Iraq war in 2003.[2] The output of the project was not a UAV, but a bidding document for the development of a UAV, including a structural design and a cost analysis.

In the figure, nodes represent activities (with expected duration in days), and arrows the precedence relationships among the activities. The critical path is the traditional notion of the project's duration (marked by the bold arrows): It is determined by the longest path from the beginning node (A1) to the end node (A10), which restricts how quickly the project can be carried out. The length of the critical path in Figure 3.2 is 57 days. The critical path is used to give a feeling for the project's duration ("we can only do it in 10 weeks if we work 7 Saturdays out of the 10 weekends") and to focus on the critical activities.

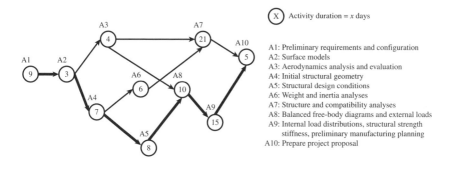

A1: Preliminary requirements and configuration
A2: Surface models
A3: Aerodynamics analysis and evaluation
A4: Initial structural geometry
A5: Structural design conditions
A6: Weight and inertia analyses
A7: Structure and compatibility analyses
A8: Balanced free-body diagrams and external loads
A9: Internal load distributions, structural strength
      stiffness, preliminary manufacturing planning
A10: Prepare project proposal

➡ Critical path

Figure 3.2   A network flow diagram of a project schedule

The problem with the critical path is that it implies that the duration is deterministic. For each activity, the "expected" durations are used. But that, of course, is fiction. In reality, activity durations are subject to variation— that is, to more or less important deviations from the expected duration. Variation is due to a myriad of little reasons that cannot be analyzed or predicted in detail, simply because there are too many of them. For instance, the engineers know that activity A1 may take between 9 and 12 days, and A7 between 18 and 27 days, and so on. The duration is not a number, but it has a (statistical) distribution.

Once we acknowledge that the activity durations have distributions, we can *simulate* the project duration.[3] Now, the project duration can be shown as a *histogram* (see the left-hand side of Figure 3.3); in other words, the project duration has a distribution, just as the activities do.

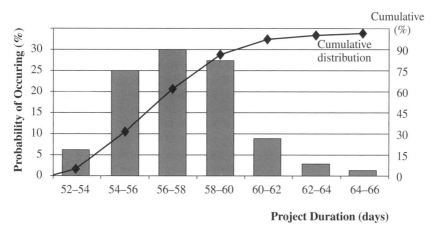

Figure 3.3   Histograms, or distributions, of the project duration

The histogram tells us several useful things: With variation, the project may be completed as quickly as in 53 days, but it may also take 65 days. Adding the bars to the right of 60 days, the histogram tells us that there is a 12 percent chance that the project may take longer than 60 days, which means that we can only get it done in 10 weeks if we work every Saturday plus several Sundays.

For the project manager, this raises the question: "Am I willing to bet that it will really take only 57 days, and schedule only 7 Saturdays, running a 12 percent risk that I will miss the bidding deadline?" In other words, making the variation explicit with simulation allows the project manager to think in terms of a service level, or a risk of missing the deadline, rather than working with a fixed estimate. If, for instance, the project manager wants to be 99 percent sure that the deadline will be made, she must schedule all the Saturdays plus four Sundays, so the deadline is missed with only a 1 percent chance that the project will take the maximum 65 days.

Project Buffers

An alternative way of explicitly acknowledging project variation, somewhat simplified in comparison to simulating the entire duration histogram, is to use project buffers. These take the form of schedule buffers, budget contingencies, or specification compromises. Buffer management has been a well-understood part of PRM for a long time.[4]

The idea is to schedule all activities at their latest start times according to classic critical path calculations (the critical path is the sequence of activities that have no "slack"—that is, for which a delay of one day immediately translates into a project delay of one day). A safety buffer is added at the *end* of the project rather than during each activity. This buffer protects the promised (deterministic) completion time from variation in the tasks on the critical path. "Feeding buffers" are placed whenever a noncritical activity feeds into the critical path, both to protect the critical path from disruptions caused by the feeding activities and to allow the critical chain activities to start early when things go well (see Figure 3.4).

A critical step is moving the "safeties" from the individual activities into the project buffer. Task completion time estimates should be at the median, implying that they are missed 50 percent of the time. As activities evolve, management keeps track of how much the buffers are consumed. As long as there is some predetermined fraction of the buffers remaining, all is assumed well; otherwise, problems are flagged or corrective action is taken. Goldratt (1997, p. 157) recommends that the project buffer be 50 percent of the sum of the safeties of the individual activities; Herroelen and Leus (2001) show that the project buffer may be even smaller, as little as 30 percent, in large projects with a "typical" structure of task distributions.

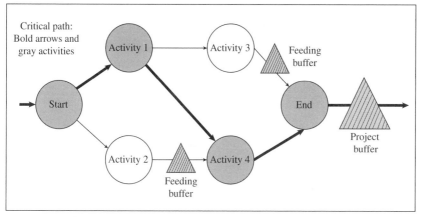

Figure 3.4 A project plan with project buffer (*Source*: Herroelen and Leus 2001).

The key to the effectiveness of the project buffer is realizing that it is not mainly a calculation device, but a tool to change attitudes: Project workers no longer need to protect their own schedule (so they no longer need to low-ball), nor can they procrastinate because they impact the overall buffer that everyone looks at and depends on. The entire team "sits in one boat." It is mostly this change in mutual commitment that has made buffer management popular over the last five years.

3.2.2 Foreseeable Events

Foreseeable events are represented by the alternative path and target in the left-hand picture of Figure 3.1. We know that certain events may take place (although we are not sure), and we can anticipate alternative courses of action that we trigger when the events occur. This is precisely the notion of risk identification and contingency planning that underlies established PRM, as described in Chapter 1, and represented by the Circored risk list in Figure 2.2.

Let us return to the UAV project described in Section 3.2.1. The project flow diagram, as depicted in Figure 3.2, is not correct. The engineers know from experience that in about 30 percent of bidding designs, the resulting load distributions and stiffness analyses indicate a problem that requires them to rerun the structural geometry, structural design, and free-body diagrams (tasks A4, A5, and A8 in Figure 3.5). Critical path analysis cannot handle loops (the path would be infinitely long, turning on itself). This inability really affects current project management tools: Try to specify a loop in Microsoft Project; it will give you an error beep!

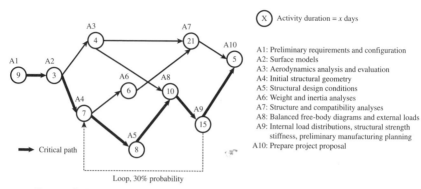

Figure 3.5 A network flow diagram of a project schedule with rework loop

It is very easy to incorporate a rerunning of activities A4-A9 with a probability of 30 percent into the simulation. The resulting histogram is shown in the right-hand side of Figure 3.6. Bad news! The entire histogram of the left-hand side of the figure is now collapsed into the far left two bars (we need a larger scale of the x-axis in order to accommodate much longer durations). If the rework loop occurs, the project duration will be between 90 and 110 days! That means there is no way the team can make the deadline, even if it works 7 days a week, and on top of that, 2 hours' (25 percent) overtime every day.

The rework loop, which is a foreseeable event, dominates the entire variation of the individual activities. If the rework loop occurs, the team might have to do something radically different. In other words, the loop goes beyond variation; it represents a major event whose occurrence is uncertain and that has an important impact on the project. The loop must be viewed and treated as a foreseeable event.

In this section, we discuss two classic approaches to foreseeable events: decision trees and risk lists.

Decision Trees

Two methods for incorporating the identified risks into the project plan are most widely used: decision trees and risk lists. We discuss both below. Figure 3.7 shows an example of a decision tree, corresponding to part of a drug research project for the development of a central nervous system drug (calcium channel receptor blocker for sleep disorder indication). Squares in the tree denote *decision nodes*, indicating decision points: Do/don't continue with the project at the stages of research, preclinical development, clinical development, and market introduction. Thus, each decision node has two branches, "yes" and "no." Under the "yes" branch, the time and cost of continuing are indicated.

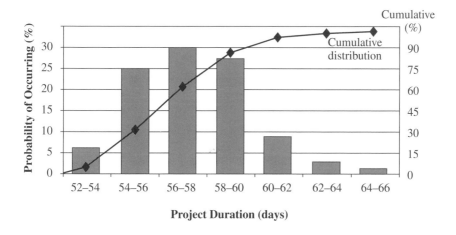

Project Duration (days)

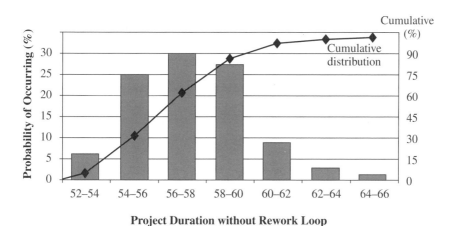

Project Duration without Rework Loop

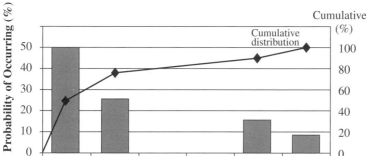

Project Duration with Rework Look

Figure 3.6   Project durations with and without rework loop

The circles in the tree denote *chance nodes*, indicating major risks: in this case, the discovery of side effects that would prevent successful market introduction of the drug. Each chance node captures one major risk, the side effects detected during one phase of the drug's development. Again, the chance nodes are simple because each one has only two branches, corresponding to "success" (the risk does not occur) and "failure" (the risk does occur). Late failure corresponds to side effects that are discovered after market introduction and force withdrawal of the drug from the market (this recently happened, for example, to Redux, a weight loss drug, and Baycol, an analgesic). The respective probabilities are indicated next to the branches (they are estimated based on historical statistics from similar drugs).

The estimated market potential of the drug is indicated on the far right, amounting to $1.8 billion in profits (not revenues!), cumulative over the life of the drug and discounted back to the time of market introduction (at an annual interest rate of 10 percent). This expected value has an estimation range of ±60%. The decision tree is analyzed backward: The value of "yes" at the decision node "market the drug?" is the expected value at the subsequent chance node, discounted by one year, minus the cost of continuing—that is, $1,466 = (.97)(1,787)/1.1 - 110$. This is higher than zero, the value of stopping, so the optimal decision is to continue. Based on the value at this decision node, the decision tree can be analyzed further backward, in the same way, up to the value of the initial decision at the root of the tree.

Accounting for the low overall success rate of 3.6 percent and the discounting over 10 years (at 10 percent p.a.), the expected NPV at the time of the first decision, if the decision is "yes," is as little as $11 million (if this value negative, it would be preferable to not engage in the project in the first place). This is typical for pharmaceutical drugs—80 percent of chemical entities entering clinical development fail, and pharmaceutical development takes a long time.

Decision Tree and Abandonment Option

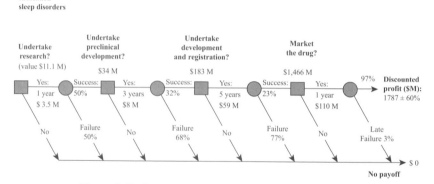

Figure 3.7   Decision tree of a central nervous system drug[5]

This example demonstrates several useful features of decision trees. First, the tree clearly identifies the *value of managerial flexibility*, or of contingent action in response to risk occurrence: If the company did not have the option of stopping after side effects occur in a given phase, all future investments would be wasted and the NPV of the project greatly reduced.[6] Second, the tree can also help to identify the *value of preventive and mitigating* action; if, for example, the failure probability after preclinical development (68 percent) could be eliminated or reduced, the value of the drug (at the initial decision) would be increased. This increase would correspond to the value of the preventive/mitigating action and could be compared to the cost of that action. Similarly, the value of additional contingent actions can be calculated; for example, in the case of a side effect, sell the drug patent for an industrial application. Third, the tree shows the *dependence among the risks*; for example, if the first one occurs, the future ones, as well as the contingent or preventive actions, become irrelevant. This dependence and ordering in time establishes a natural order of attention for the project manager.

Thus, a decision tree is a powerful tool for risk identification, a tool that not only identifies risks but also facilitates the subsequent PRM phases of risk prioritization and risk management. A decision tree offers a way of looking at project risks in a conceptually clear framework. However, decision trees have an important drawback: Their complexity explodes exponentially with the number of risks and decisions considered (for each decision and risk with n branches, the number of subsequent subtrees is multiplied by a factor of n). Even when it might still be possible to "crunch the numbers" of the tree on a powerful computer, the data-gathering effort quickly becomes unmanageable and the result of the tree analysis intransparent, and therefore much less useful, for the decision making team or manager.

The exponential explosion renders decision trees unusable for projects with large numbers of risks. Therefore, decision trees are commonly used only to *focus* on a handful of the most important risks. Sophisticated project management companies—engineering service providers, for example— perform this focused analysis, ignoring other "smaller" risks at the first cut and then incorporating them through risk lists. The pharmaceutical industry uses decision trees extensively in this way, which is facilitated by the fact that the *effect of major risks is simple*, that is, decision and chance nodes have only two branches (*go/kill*), and thus a relatively large number of risks can be incorporated without losing transparency.

Risk Lists
A risk list is a simpler tool than a decision tree. It simply describes each risk separately, with its nature, its effect, its probability, and preventive, mitigating, or contingent actions. Unlike decision trees, risk lists do not explode in complexity when the number of risks is large. If the risks do interact (that is, if a downstream risk looks different, as a result of what happened upstream), the simplification loses information compared to the decision tree. However, many project risks have a "local" effect; they do

not influence the actions downstream. In this case, a risk list is fully adequate, and a decision tree is not necessary at all.

Another advantage of risk lists is that they can be summarized in "generic" templates that group all the risks that have occurred in the past, but without the actual numbers (impact, probability). Such templates are a powerful way of summarizing experience (as we have seen in the PCNet project). Figure 3.8 shows a summary of a generic risk template from the pharmaceutical industry, analogous to the risk list that was used in the PCNet project. The full template is 20 pages long; it embodies experience about risks in pharmaceutical development.

3.2.3 Residual Risk

Residual risk is depicted in the top picture of Figure 3.1 as the response to the hostile wind that threatens to blow the project off course, but does not require a fundamentally different approach. Residual risk is what is left over after planning for foreseeable uncertainty. In many projects, there are simply too many foreseeable events, and planning for each event becomes impossible. While many of these events, if small enough, may be captured in the project variation, some may have quite large impacts on the project.

For example, when HP merged with COMPAQ, it ran into major trouble when it migrated its industry-standard servers (ISS) division to an SAP enterprise resource planning (ERP) system. While HP had invested significantly in contingency planning, not only on the IT side but also by putting aside inventory and capacity in order to mitigate against any IT disruptions in the customer order process, it did not anticipate the extent to which the migration could disrupt its business. As Gilles Bouchard, CIO and EVP of global operations for HP, was quoted as saying, "We had a series of small problems, none of which would have been too much to handle, but together they created the perfect storm."[7] When this "perfect storm" hit, as many as 20 percent of HP's customer orders were kicked out of the ERP system because of its inability to deal with certain customized orders. The contingent inventory and capacity put aside by HP to mitigate against any disruption proved insufficient to cope with this level of disruption. HP estimates that it lost $40 million in revenue: $10 million more than the cost of the entire IT project.[8]

This is the dark side of contingency planning. Weick and Sutcliffe call this danger of contingency planning "double-blind": "Contingent actions are doubly blind. They are blind because they restrict attention to what we expect, and they are blind because they limit our present view of our capabilities to those we now have. When we plan contingent actions, we tend not to imagine how we might recombine the actions in our current repertoire to deal with the unexpected. In other words, contingency plans reduce improvisation."[9] Of course, the managers of the PCNet project in Chapter 1 may retort that they planned thoroughly *and* improvised with the help of the PRM office. The question is not one of either/or, but how to do both.

Risk Category	Detailed Subcategories
Substance and Production	
Ingredients	Risk from suppliers (dependency, stability, transfer, contracts), cost of production, availability of drug substance, process (reproducibility, scale-up, impurities), stability (shelf life)
Final product	As above, plus dosage changes, formulation changes
Analytical methods	Specificity, transfer of license or to a different site
Regulatory issues	Ingredient status, toxicity documentation, mixtures, impurity limits
Preclinical	
Safety pharmacology	Findings in core battery studies, supplemental studies, toxicity in cell cultures
Primary pharmacology	Choice of endpoints and species, target selectivity, and specificity
Bioanalytics	Detection of parent compound and metabolites, toxicity or metabolism in test species different from humans, drug accumulation, oral bioactivity, in vivo tests, body penetration
Toxicology	Availability of test substance, pharmacodynamic side effects, high mortality rate in long-term studies, drug-specific side effects
Clinical	
Phase I	Pharmacokinetics (e.g., different in subpopulations, interactions with other compounds or foods), pharmacodynamics (e.g., subject tolerance different from patient tolerance)
Phase II	Appropriate dosage, exposure duration, relevance of placebo control
Phase III	Study delay (e.g., because of season), patient recruitment (e.g., tough criteria, special patient groups, dropout rates), negative outcome (not significant), new regulatory requirements
General Regulatory Risks	Status of comparator, toxants in environment, availability of guidelines, interaction with agencies (e.g., process time, contradictions among different agencies), requirement differences across countries
General Risks	
Licenses	Dependence on licensing partners
Patents	Disclosure of new patents
Trademarks	Viability/acceptance of trademark at submission
Costs	Currencies, inflation, additional patients or studies needed
Market risks	New competitors, new therapies, patient acceptance, target profile, political risks (e.g., pricing, prevention versus therapy)

Figure 3-8   Generic risk list (template) of a pharmaceutical development project

Organizations must recognize that no amount of contingency planning will identify all the risks or all the combinations of foreseeable events that might happen. Thus, organizations must be prepared to deal with them as they arise. As there is no contingency plan for dealing with such events, the project team must be prepared to improvise and respond quickly to events as they arise. We saw in the PCNet project in Chapter 1 how a separate team and dedicated resources were assembled to deal with the residual risks.

3.3 Complexity

The contingency planning approach is based on the obvious and fundamental assumption that the project and its contingencies can be planned. In other words, a near-best set of actions can be specified for any course of foreseeable events. Many project management tools, such as the critical path method, PERT (Program Evaluation and Review Technique), GERT

(Graphical Evaluation and Review Technique), and Q-GERT have been developed to assist project teams in finding these near-best courses of action. However, many projects are simply too complex to yield a near-best solution—that is, they are simply unpredictable.

Complexity stems from "a large number of parts that interact in non-simple ways [such that] given the properties of the parts and the laws of their interactions, it is not a trivial matter to infer the properties of the whole."[10] Complexity has two "ingredients": system size (the number of parts) and the number of interactions among the parts. A large system is not complex if the parts do not interact—we can treat them in isolation and simply add them up to understand system performance. The essence is that the interactions of the elements make the system more than the sum of the parts and the system's behavior hard to predict from the behavior of the parts.[11] Complexity makes it difficult to find the "best" configuration of a system. Project complexity is a typical challenge in many large-scale projects: Too many combinations of actions and influence variables exist, all with different performance implications that cannot be extrapolated from similar combinations. In this case, it is impossible to "optimize" the project plan, and the team has to find an acceptable plan that works satisfactorily.

This was certainly the case in the Circored project—there were over 300 dynamic flow control parameters alone, and decisions on temperature and flow rate interacted with material characteristics (such as viscosity and abrasiveness) and equipment specifications (such as stiffness, toughness, and brittleness).

We can take another example from the car industry, which is currently experiencing an intensive discussion of future hydrogen cars. A hydrogen engine car "system" can be viewed as being made up of the "chunks" engine (including hydrogen storing tank and injection under high pressure), drive train, chassis (axles, wheels, brakes, frame), exterior body, passenger compartment, driver interface and controls, market positioning, and infrastructure (permission and hydrogen refueling infrastructure). Each chunk consists of many components.

Note how market and infrastructure are viewed as chunks of the "system," as they influence success, pose uncertainty, and interact with the design of the physical product. Engine, chassis, drive train, and body also interact among one another because of physical dimensions, exchange of forces, and mutual impact through control systems. Note also that the number of interactions may include not only technical interactions (task complexity), but also interactions among interests of multiple stakeholders (relationship complexity). This system contains so many interactions with so many performance "peaks" that it cannot be completely understood. The design goal is something that works; "optimization" is elusive.[12]

The network planning approach (critical path), shown in Section 3.2, can deal with some degree of complexity: the number of interactions among many activities that are caused by precedence relationships. This level of complexity is (most often) rather low because any activity has precedence relationships (at least ones that influence the plan) with only a few other activities, so the number of interdependencies is small. We discuss the diagnosis of complexity in a project further in Chapter 4.

To deal with high complexity, we discuss two existing methods: control-and-fast-response, or "high-reliability organizations" to deal with task complexity, and project contracts for relationship complexity.

3.3.1 Task Complexity: Control-and-Fast-Response

Control-and-fast-response is a useful and interesting approach to risks that follows a different mind-set than established PRM. In their book, *Managing the Unexpected*, Karl Weick and Kathleen Sutcliffe discuss what high-reliability organizations, such as a nuclear power plant or an aircraft carrier, must do to guarantee a reliable functioning of a very complex system. Reliable operation must be guaranteed (almost at all cost) because much is at stake.

A striking example is the operation of an aircraft carrier: " . . . you have six thousand people crammed into tight spaces away from the shore on a 1,100-foot, 95,000-ton floating city run by an overburdened 'major.' Within those tight spaces on a carrier, you also have people working with jet aircraft, jet fuel, nuclear reactors, nuclear weapons, an onboard air traffic control system, refueling and re-supply from adjacent ships that are moving, a surrounding battle group of seven to nine ships that are supposed to protect the carrier but that can themselves also be dangerous obstacles in fog or high seas and unpredictable weather. The list of 'gee whiz' stuff on a carrier seems endless."[13]

People on a carrier cannot afford to be wrong, or lives will be lost. This is a huge challenge because the system is so complex—the different parts of the carrier are tightly coupled, and impact one another, and the individual components constantly change, because, for example, of human error, equipment failure, or changing weather conditions. "Safety is elusive because it is a dynamic non-event—what produces the stable outcome is constant change rather than continuous repetition. To achieve this stability, a change in one system parameter must be compensated for by a change in other parameters."[14] And yet, accidents rarely happen.

Weick and Sutcliffe recommend that the organization develop what they call "mindfulness." This refers to "the combination of ongoing scrutiny of existing expectations, continuous refinement and differentiation of expectations based on newer experiences, willingness and capability to invent new expectations that make sense of unprecedented events, [and] a more nuanced appreciation of context and ways to deal with it."[15]

Mindfulness includes a number of "soft skills," such as preoccupation with failure, reluctance to simplify, sensitivity to operations, commitment to resilience, and deference to expertise. In our language of "systems," mindfulness means the ability to know precisely what the "in control" target state of each component of the system is, to detect even small deviations from the target state, and to quickly react to them and contain them so that they do not spread to other components of the system, causing a major problem there. In other words, mindfulness represents *control-and fast-response*: We prevent deviations if possible, and if one occurs, we contain it immediately. We will come back to mindfulness in Chapter 8, when we discuss the project mind-set that prepares a team for unk unks.

Control takes the form of preoccupation with failure, or ever-paranoid and pervasive monitoring. For example, aircraft carriers conduct foreign-object-damage walk-downs on deck several times a day to prevent small objects (such as bolts or trash) from being sucked into airplane engines. In the constant chatter of simultaneous loops of conversation and verification, "seasoned personnel do not 'listen' so much as they monitor for deviations, reacting instantly to anything that does not fit their expectations of the correct routine."[16] Similarly, high-performing nuclear power plants conduct almost daily departmental incident reviews of seemingly minor slips that have no obvious link to any consequential damage.[17] Reluctance to simplify means that deviations are not conveniently explained away but investigated until their root cause is found. Deference to expertise refers to the principle that the judgment of the people who know the daily operations is respected, even if it is disruptive or painful, no matter where they are in the hierarchy.

When a slight deviation is discovered, even if it seems inconsequential, corrective and, if necessary, drastic action is taken. For example, a seaman on the nuclear carrier *Carl Vinson* reported the loss of a tool on the deck. All aircraft aloft were redirected to land bases until the tool was found, and the seaman was commended for his action—recognizing a potential danger—the next day at a formal ceremony.[18] Commitment to resilience means the ability to substantially deviate from established routines, and to modify those routines, in order to mitigate the deviations before they escalate out of control.

In Section 1.7, we contrasted the critical path mentality with the PRM mentality: the assumption that there is a well-defined target, a path to get there, and the project must reach it versus the prevision of the need for alternative targets and paths, and a willingness to switch to contingencies. Control-and-fast-response embodies yet a different mentality: It admits that there is a wide "state space" of influence factor configurations out there, which contains many nasty surprises, and therefore we insulate the system from this state space and keep it iron-fisted at the state that we know works. Figure 3.9 repeats the center pane of Figure 3.1.

Compared to PRM, the emphasis is not on planning contingencies but on mutual adjustment of the system elements (such as ground crew, pilots, and ship operations) to bad news that emanates from different system elements, in order to keep the system in the control state, or to minimize deviations from it before they escalate. This relies not only on planned routines but also, critically, on a willingness to improvise (resilience) if that particular combination of circumstances has not been foreseen. And because of system complexity, it is not possible to anticipate all system constellations.

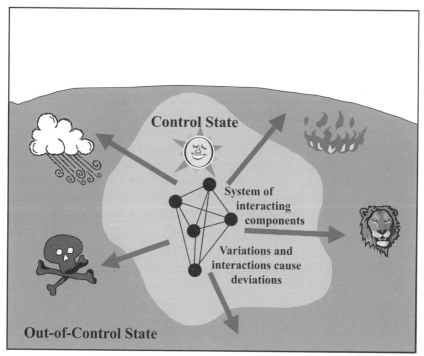

Figure 3.9   The control-and-fast-response mind-set

Control-and-fast-response and mindfulness are highly relevant to project management for two reasons. First, they provide a good discipline of knowing as much as possible and reacting to deviations that are not required for learning about the path toward the goal. Second, mindfulness helps to alert us to the problems of complexity, the interactions among multiple system parts, as a major source of risks. Mutual adjustment and resilience are highly applicable in project management.

3.3.2 Relationship Complexity: Contracts as Risk-Sharing Tools

Complex interactions arise not only from the interdependence of the tasks, but also from conflicts of interest, stemming from relationships among the parties that are involved in a project. We refer to this as relationship complexity. This is becoming more and more relevant as project participants increasingly come from different organizations. A widely used tool for defining the interests of the players and for sharing risks is a *project contract*.

Large projects are rarely performed with one organization's internal resources alone: The resource commitment is too great, the risk becomes too high, and the range of specialized expertise areas goes beyond what exists in one company. Therefore, managing major projects typically involves working with partners. Collaboration with external parties poses a trade-off—the above advantages have to be weighed against multiple interests, which are never perfectly aligned and which cause possible interactions among multiple influences, or, in other words, complexity.

Contracts are the most widely used way of handling external partners. Contract management has developed its own jargon, but in this section, we argue that the logic behind contract design is very closely related to PRM. Project management literature distinguishes three major contract forms: fixed price, cost reimbursable, and mixed incentive contracts, summarized in Figure 3.10.[19] They differ in their appropriateness in allocating risks. Lump-sum turn-key (LSTK) fixed-price contracts allocate total risk to the contractor; they seem to have increased in importance over the years, as they clearly allocate responsibility to one major contractor who assumes most risk and can control the project's execution, minimizing interfaces and working with more overlap.

In cost-reimbursable contracts, including engineering, procurement, and construction management (EPCM) contracts, the contractor is entitled to charge all justified costs. The client must drive the project, investing more resources and assuming all risks. The contractor has little incentive to be efficient. There are also "intermediate" contract types, involving incentive fees, bonuses/penalties, and target prices to compel the contractor to trim costs without sacrificing quality. These contracts are not used as widely as one might expect, because negotiating targets is complex, and because the required implementation involvement by the client is (almost) as high as in a cost-reimbursable contract.[20]

Although contracts are agreements among partners, they must include elements of a "hierarchy" (as if the parties were coordinated internally within one organization), in order to be operable during the myriad of small decisions to be taken during execution. These elements of hierarchy include command structures and authority systems, dispute resolution procedures, standard operating procedures (SOPs), and incentive systems. In other words, these hierarchical elements are necessary in order to respond to the residual risks that we have seen in the PCNet example.

A useful view of a contract is that of a *business deal.*[21] The contract must, therefore, above all address the major contents of the business proposed. Specifications define the business function of the project outcome, and price and schedule the investment, with payment terms determining the timing. Then there are multiple tools for mutual insurance, warranties, damages and limitations to them, and securities. They are depicted in Figure 3.10 and further explained in Table 3.1.

The contract shapes the culture: first, because a project is not a permanent relationship in which the prospect of future interaction would discipline behavior, and second, because personnel turnover during the project is common. Thus, the contract is the key framework for setting standards of behavior and trust shown by others and, ultimately, the project's performance. The perceived fairness, realism, completeness, and transparency of the business deal are key elements in building up needed trust.

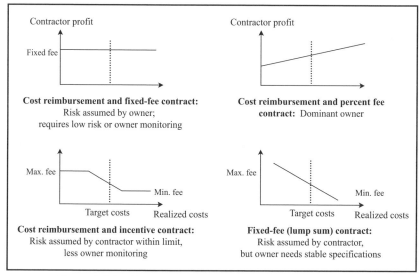

Figure 3.10 Contract types and risk allocation among the parties

Content of the Project

- Fulfilling the specifications
- Within budget price/cost estimates
- According to schedule
- Payment of the contract price, with payment terms

Assurance for both sides:

- *Warranties* (faults after delivery)
- *Liquidated damages* (penalties for nonperformance)
- *Limitation of liability* to protect contractor
- Mutual assurance of fulfilment with *securities*

Figure 3.11 The eight key business levers in the contract

As the contract sets the tone of the collaboration, it is critical that the price in Figure 3.11 is based on reasonable cost estimates for the project. While the price is a zero-sum game in the short term (the client wants to get the best deal while the contractor wants to make a living), deviating from the true cost in either direction is very dangerous: If the price is too low, the contractor will feel an irresistible temptation to shirk (there is no complex project in which the contractor cannot save costs by compromising on quality). If the price is too high, the client may not react this time but may find out and retaliate next time. Either side should avoid dictating contract terms and conditions, no matter how powerful he is; virtually always, *both* sides have the opportunity of shirking. This implies that contracts, especially fixed-price LSTK contracts, should not be awarded on the basis of the lowest bid but based on identified risks, capabilities, and track records.[22]

In other words, the eight business levers represent the basis for risk identification—they are areas of high risk impact, and each area should be underpinned by careful estimates, which represent nothing other than PRM risk identification and assessment. While the eight key business drivers determine the fundamental logic of the business deal, the additional hierarchical contract components govern the micro-interactions during the project. The hierarchical contract components manage inevitable residual risks. For example, the parties must agree upon a process via intermediate deliverable deadlines such as document sign-off, equipment inspection and triggering of payments, change-order procedures, and conflict resolution.

Table 3.1: Definition of the Key Drivers of a Contract Business Deal

Key Driver	Definition/Key Issue to Be Clarified
Technical specifications	Adequacy, completeness, and consistency of the description of the scope of work. Consistency between technical and commercial parts.
Price (quality of cost estimates)	Consistency of price and cost estimates with technical specifications. Adequacy of contingency and profit margin.
Payment terms	Schedule of partial payments. This determines to what extent cash receipts by the contractor cover his cash expenses over the course of the project, defining the contractor's exposure from cash flow during the project.
Schedule	Achievability of key (intermediate and final) completion dates and consistency of their definitions. Impact of possible project delay/acceleration costs relative to contractual liquidated damages.
Performance guarantees	Acceptable tolerances of key performance measures; definition of preconditions for achievement of these performances; and liquidated damages that compensate for deviations from the performance tolerances.
Warranties	Payments for the repair or replacement of unsuitable or defective equipment. Possible compensation for consequences of defective services (such as engineering).
Limitation of liability	What is the maximum extent of the contractor's liability toward the client under the contract (excluding tort or negligence)? Is it contractually clearly limited, and are indirect and consequential damages excluded?
Securities	How does the contractor ensure his performance toward the client? How does the client ensure his payment obligations toward the contractor? For example, deposits, bonds, or guarantees by third parties.

In summary, project contracts are a widely used way of governing projects with multiple parties, by implementing PRM across different organizations. While the language used in contract design is different from the language used in PRM, the principles of PRM (risk identification through key business levers, monitoring, and hierarchical structures to respond to residual risk) are consistent with contract design principles.

3.4 Unknown Unknowns

We have already discussed in this chapter, and elsewhere, that in major projects, not all project influence factors can be foreseen and planned for—some of them are not known by the project team at all. The same effect results if the project team is not aware of major interactions among influence variables and actions. They are not within the team's horizon; they are outside its knowledge. Therefore, the team cannot plan for them. In addition, there are actions (relating to these unknown variables) of which the team is not aware. The decision theory and economics disciplines call this "unawareness" or "incomplete state space," and technology management scholars call it "ambiguity." As mentioned, in project management, unforeseeable uncertainty has been referred to as unknown unknowns, or unk unks.[23] Weick and Sutcliffe call unk unks "bolts from the blue," referring to events for which the team had no expectation at all, no hint, and no prior model.[24]

Unk unks are fundamental for novel projects. This has been acknowledged by experts before. For example, Miller and Lessard conclude that the challenge is "ignorance of the true state of nature and the causal structures of decision issues."[25] Similarly, researchers of new venture startup projects have observed, "What has made or broken the companies . . . is the ability or inability to recognize and react to the completely unpredictable."[26]

We have already discussed one response to unk unks in residual risk management in Section 3.2.3. Managing residual risk implies first a recognition that there are things you do not know that you do not know (unk unks), and second the ability to improvise to either take an alternative path or to bring the project back to the known path. Residual risk management works when the basic control state is known—that is, we have a pretty good idea what we want to do. For example, none of the residual risks in the PCNet project fundamentally changed the nature of the project: It was known what they wanted to achieve and how they would achieve it, more or less, and residual risk management mostly dealt with events that threatened to blow the project off course.

The second approach to unk unks that we have discussed is control-and-fast-response (Section 3.3.1). The idea is to avoid the unk unks altogether by an instantaneous and iron-fisted reaction to any deviations from the control state of the project before they can spiral out of control.

Control-and-fast-response works if a well-defined target state of the project can be identified and if actions are available to maintain it.

However, unk unks may be so fundamental that the project goal and path are, themselves, fundamentally unknown. There is no path from which residual risk management can deviate; there is no control state to be maintained. This is the situation that is represented in the right-hand picture of Figure 3.1, and that was faced by the Circored project: It was unknown what part of the scale-up would not work and what type of actions might be required to make the chemical process work.

If a project is faced with important unk unks of this type, there is really no project plan. Or, any plan is a fiction. It corresponds to a set of stakes in the desert (Figure 3.1), of which we do not know whether they lead to an oasis or not. Any project plan will run into major surprises (many of them negative), and the plan will miss major actions that arise as attractive ex post but were not identified ex ante. A serious danger is that the stakes in the ground are easily interpreted as a real plan, as if they "claimed" to be a plan. Planning is always necessary in order to have a base line; however, a plan in the presence of unk unks may constitute "false precision," misleading a project team to be less alert to changes than is required.

We must plan; it is the basic building block on which everything else rests. However, in novel projects, where we face fundamental knowledge gaps and must be ready for unk unks, adherence to a plan (even with contingencies) must not become an end in itself. Unk unks (whether they come from unknown influence factors or from complexity and ill-understood interactions) require the readiness to abandon assumptions and look for solutions in nonanticipated places.

In the presence of fundamental unk unks (see the right-hand picture of Figure 3.1), flexibility in dealing with residual uncertainty will not be sufficient to respond to major unk unks that were not visible at the outset and emerge only mid-course during the project. Nor can a control-and-fast-response attempt to maintain a control state that cannot be defined. This is what the Circored project experienced, and what damaged the careers of several of the people involved.

At this point, it is worth coming back to complexity (Section 3.3). Not only does complexity prevent a project team from designing the "optimal" project plan, but it is also a major source of unk unks if the nature of the interactions among the project influences and actions are not fully known or understood: If there is a slight deviation from the plan in one variable (let's say, process temperature in the Circored facility), it may cause major problems in another part of the project (for example, difficult forming and sticking in the briquetting machine). Moreover, these problems cannot be foreseen by the people responsible for that part of the project because the interaction was not understood. Nonanticipated interactions often cause major crises in large engineering projects and require an addition to the classic project management toolbox.[27]

Unstructured problems that are not amenable to a planned solution approach have long been identified in the design community as "wicked problems."[28] Coming from a background of urban planning and policy making, the design community characterized wicked problems as those that do not have a definite formulation, have no stopping rule that allows one to determine when the problem is solved, where solutions cannot be fully tested and the problems cannot be generalized, and where there is ambiguity about problem causes. In other words, these are ill-understood problems with major unk unks. Wicked problems are the opposite of "tame" problems that we know how to solve in science and with formal project management methods.

3.4.1 Two Fundamental Approaches: Learning and Selectionism

Figure 3.12 outlines the logic of three fundamental approaches to project management in the face of the different sources of risk. The first (Figure 3.12a) is, again as a reminder, a summary of PRM, as described in Chapter 1: If we have an adequate picture of the influence variables, the causal effects of our actions, and the resulting performance, we can choose a desired outcome and a course of action that is manifested in a project plan. We may plan for contingencies if new information about a (initially identified) major risk emerges, and we may have to "improvise" around the target outcome in order to respond to residual risk. But fundamentally, the approach is a *planned approach*; the important problem solving occurs at the beginning and then the emphasis shifts to executing the plan.

We have seen in the Circored example that this approach leads to negative surprises and crises if the unk unks are so significant that residual management around the target outcome does not adequately address them. Figures 3.12b and 3.12c show two fundamental approaches that project teams can use to respond to major unforeseen influences.[29]

If we admit that we know too little about the universe of possible project outcomes (and how to get there), we may not insist on choosing a target outcome at the outset. Nor may we try to maintain a control state, because the target state is unknown; this is the limit of the applicability of control-and-fast-response to project management. Rather, we start moving toward one outcome (the best we can identify), but we are prepared to repeatedly and fundamentally change both the outcome and the course of action as we proceed, and as new information becomes available. In other words, we *iterate and learn* (Figure 3.12b). The most important problem solving is distributed at the outset and throughout the duration of the project.

This approach has been given different names by previous project management workers. For example, Chew et al. (1991) examined unk unks in the context of introducing new manufacturing technologies in plants and concluded that iteration, learning, original new problem solving, and adjustment are required. In the context of new product development,

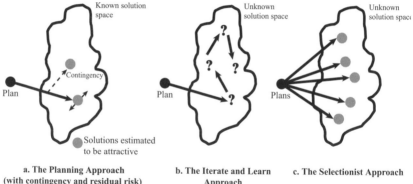

a. The Planning Approach
(with contingency and residual risk)

b. The Iterate and Learn
Approach

c. The Selectionist Approach

Figure 3.12   Three fundamental PRM approaches in face of uncertainty

Leonard-Barton (1995) called the iterate-and-learn approach "product morphing" (meaning repeated changes of a product concept over time), and Lynn et al. (1996) called it "probe-and-learn," referring to repeatedly pushing a project all the way into the market and then iterating *after* market introduction. In general, iteration and experimentation are a fundamental feature of problem solving in innovation and engineering projects[30] as well as venture startup projects.[31] It cannot be overstated that this is difficult to do—it feels uncomfortable (especially to senior managers) not to have the feeling of control that stems from defined targets, and repeated iterations are time-consuming and expensive.

The other fundamental approach is to try out several plans and see ex post what works best (Figure 3.12c). Again, this approach has been identified before—in operations research and engineering (addressing the solution methods for very complex problems), it is called "parallel trials," and in management, Leonard-Barton (1995) has called it "Darwinian selection," and McGrath (2001) has called it "creating requisite variety" for the complex problems to be solved by the organization. We emphasize the "selectionist" logic because the fundamental feature is that one out of many trials is selected ex post (whether the trials are executed in parallel or one after the other is secondary). Again, this is difficult to do—executing multiple parallel attempts is expensive, and the parallel teams may compete rather than collaborate if everyone knows that only one team will be chosen in the end.

3.5 Expanding the Toolbox: Fundamental Approaches to Project Uncertainty

The sources of project uncertainty are placed in relation to one another in a framework in Figure 3.13. The vertical axis represents low and high complexity. The horizontal axis does not represent a cardinal measure of "more" or "less" uncertainty; it simply shows the different sources, namely variation, foreseeable influences, and unknown unknowns. It is important to remember that several of these can be present in a project at the same time.

We discussed in Chapter 1 how PRM methods can deal with variation and foreseeable influences. The project management community also possesses powerful methods for dealing with complexity: The traditional network planning methods (such as PERT and CPM) are designed to schedule and control many (up to thousands of) activities with many predecessor interactions. In addition, we know that high complexity requires tight coordination and frequent communication among the parties that are responsible for the many interacting project tasks: Changes in any one task may propagate to other tasks and cycle around, and therefore, the fast exchange of preliminary information is of paramount importance. Some of these coordination principles were popularized under the topic of "concurrent engineering" during the 1990s.[32]

Our methods are less well developed in the face of unforeseeable influences, or of foreseeable influences that heavily interact, so a foreseen change in one influence factor may ricochet around and cause unforeseeable changes in other task teams or system components.

The Circored team at CAL also faced such interaction problems: Many of the component problems might have been foreseeable, in principle, through a dedicated analysis, but there were simply "too many trees in the forest," as the manager expressed it. In a novel project such as Circored, control-and-fast-response is not feasible because the "allowable operating regime" is not yet established. Yet, it is important to understand this approach because it establishes an ideal and because it alerts us to the danger of the combination of complexity and (foreseeable or unforeseeable) uncertainty. Unk unks may reflect true unknown influences, or they may arise from ill-understood interactions among, in principle, known influences.

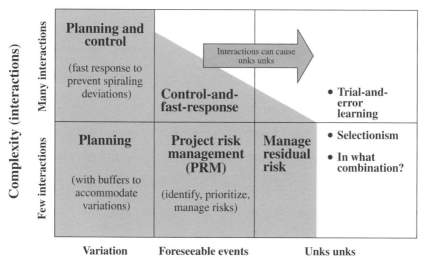

Figure 3.13 A framework of the sources of uncertainty in project management

The obvious question that we want to answer is this: What can project managers do in order to respond to complexity and unk unks in a way that avoids the negative experience of a Circored project? In this chapter, we outline two fundamental approaches. The rest of the book examines how these two approaches can be put into practice.

As we mentioned above, these fundamental approaches have been identified before, and both are used in practice. Even the field of business strategy has undergone an evolution over the last decade that is parallel to what we are proposing here: Strategy has moved from emphasizing planned and contingency approaches to "emergent strategy" that changes over time in unforeseeable ways.[33]

However, selectionism and learning are often used ad hoc and piecemeal, not as parts of an overarching strategy and toolbox for dealing with unk unks. No conceptual map and toolbox exist that compare the relative strengths and weaknesses of iteration and learning versus selectionist trials, nor do we understand well how they can fruitfully be combined. It is these managerial challenges for which this book proposes solutions.

Are There More Project Management Approaches than These Three?

We have discussed three fundamental approaches to cope with uncertainty: planning and triggering contingencies, selectionism, and learning. The reader may ask whether this representation is complete or whether other approaches exist. We have conducted a large number of case studies, and we have not found any other strategy to cope with uncertainty and complexity. Nor has any approach that is conceptually different been reported in academic and managerial publications on project management and product development (obviously, detailed implementations always differ). That is, of course, not proof that there would not be more approaches. We have, however, been able to demonstrate by a comparison with work in biology that the set of the three approaches is complete and robust.[34]

The project management challenges of dealing with uncertainty is very similar to the one that is known in biology as the "uncertain futures problem," or how to accomplish the successful propagation of a species into the next generation.[35] Plotkin has indeed shown that nature has three and only three responses. One strategy for a species is to avoid uncertainty by restricting itself to ecological niches that are simple and change slowly. However, such a strategy can be devastating if there are sudden changes in the environment. A more flexible approach of a "planned approach," with the ability to cope with foreseeable uncertainty, takes the form of contingent policies. For example, many species tolerate variations in their physical state (e.g., body temperature, caloric intake) up to a certain degree and "genetically trigger" adjustments when this variation exceeds a certain threshold (e.g., growing fur in the winter). This is a very similar approach to planned projects with contingency plans.

Some species have an ability to learn to adjust to their environment. They have the ability to extend their behavior beyond prespecified triggers by perceiving critical new features of their environment and replanning, or modifying their behavior accordingly. One example of this is the reaction of immune systems with pathogens. Biologists have shown that these learning devices are metabolically costly and that only a limited number of species have developed a learning capability.

Certain species have no ability to learn yet have a tremendous ability to adapt in a new generation to new environments that lie outside their historical experience. As each individual offspring dips into the gene pool, coming up with variants of genetic instructions, the resulting genetic variation increases the chance that some will survive. For example, bacteria with fast propagation and high mutation rates have conquered niches that were, until recently, believed to be hostile to life forms (e.g., hot sulfur vents in deep seas).

These three strategies used by nature to cope with uncertainty are conceptually similar to the project management approaches we have described in this chapter. Evolution provides an unparalleled database of strategies to deal with uncertainty and complexity. If nature, with its over 3 billion years of creative solutions, has produced the same three fundamental strategies that we find, this is corroborative evidence that there are no other fundamental strategies.

Endnotes

1. See, for example, Morris and Hugh 1987, Schrader et al. 1993, Hamel and Prahalad 1994, Miller and Lessard 2000, or Pich et al. 2002.

2. This example is based on Loch, Kavadias and De Meyer 2000.

3. That means that we draw each activity duration randomly, using its distribution, and calculate the critical path for those durations (the project duration). Then we draw a different set of random durations and calculate the critical path again, and so on, thousands of times. The thousands of project durations give a probability distribution, shown as a histogram. This can be easily done on the computer, even using simple tools such as Excel, but professional project planning packages have the simulation capability built in, for example, the Graphical Evaluation and Review Technique (GERT), for which commercial packages exist. A positive feature of simulations is that they are usually very robust with respect to the precise distributions of the activity durations. In other words, as long as we get the expected values, and the minima and the maxima of the activity durations roughly right, the histogram will be in the right ballpark.

4. See, for example, Goldratt 1997, or Herroelen and Leus 2001. Buffer scheduling exists as an add-on to commercial scheduling software packages. See an overview in Herroelen 2005.

5. Source: Loch and Bode-Greuel 2001.

6. This seems obvious in this simple tree, where each decision has only two branches, but the existence of managerial flexibility is much less obvious and, in fact, is often overlooked in projects with more complicated multibranch decisions.

7. See "When Bad Things Happen to Good Projects," *CIO Magazine*, December 1, 2004.

8. Ibid.

9. Weick and Sutcliffe 2001, p. 80.

10. Simon 1969, p. 195.

11. See also Sommer and Loch 2004, and Williams 2002, p. 50.

12. Our definition of complexity is consistent with that of other authors, although the terminology differs. For example, Shenhar (2001) calls a large project that combines task complexity and relational complexity an "array." Williams (2002) emphasizes the different interdependencies among system components (sequential, reciprocal, pooled), and he views *uncertainty as an aspect of complexity*. We believe that it is important to distinguish the two concepts, as their fundamental effects are different—complexity causes many local performance peaks in decision space, making the search for the best system solution difficult. Uncertainty, in contrast, makes the performance landscape "shift under your feet." We emphasize several times in the remainder of the book that complexity can cause uncertainty (even unforeseen uncertainty) for subprojects, or parties in the project, if the different parts of the project do not coordinate. However, this does not make uncertainty an aspect of complexity; it is an additional concept (lack of coordination) that connects complexity and uncertainty.

13. Weick and Sutcliffe 2001, p. 28.

14. Weick and Sutcliffe 2001, p. 30.

15. Weick and Sutcliffe 2001, p. 42.

16. Weick and Sutcliffe 2001, p. 32.

17. Weick and Sutcliffe 2001, p. 57.

18. Weick and Sutcliffe 2001, p. 59.

19. For example, Stinchcombe and Heimer 1985, Kerzner 2003, Ferreira and Rogerson 1999.

20. See Ward and Chapman 1994.

21. This is proposed by Von Branconi and Loch 2004.

22. See Hackney 1965, Chapman and Ward 1997, and Von Branconi and Loch 2004.

23. An example of technology management work that introduced the term "ambiguity" is Schrader et al. 1993. For the term "unk unks," see, for example, Wideman 1992, although the term has been used in aerospace and electrical and nuclear engineering for decades. Floricel and Miller 1998 call the unk unks "strategic surprises."

24. Weick and Sutcliffe 2001, p. 36.

25. Miller and Lessard 2000, p. 76.

26. See Brokaw 1991, p. 54.

27. Miller and Lessard, 2000, have demonstrated the effect of unanticipated interactions. Williams, 1999, calls for new project management approaches for complex projects.

28. This term was coined by Rittel and Webber 1973.

29. Why are we proposing just these two approaches? Examinations of what has been proposed by previous analyses, as well as theoretical considerations, suggest that all responses to unforeseeable uncertainty represent combinations of the ones we discuss here (see Pich et al. 2002). See also Box on p. 77.

30. See, for example, Van de Ven et al. 1999 (Chapter 2), or Thomke 2003.

31. See, for example, Drucker 1985, Pitt and Kannemeyer 2000, or Chesbrough and Rosenbloom 2003.

32. For overviews, see Mihm and Loch 2004, Terwiesch et al. 2002, Smith 1997.

33. From a project management perspective, see the discussion in Boddy 2002, pp. 49–52. In the strategy field, see, for example, Mintzberg 1994, who was one of the first to call for a new paradigm in strategy, Bettis and Hitt 1995, or Adner and Levinthal 2004. The latter authors argue that real options evaluation (which is equivalent to decision trees when projects cannot be hedged in financial markets) is not enough in strategy when target markets are highly uncertain.

34. Pich et al. 2002.

35. Plotkin 1993, pp.145–48.

PART

II

Managing the Unknown

Part I of this book gave an overview of established PRM and showed the limitations of this approach when unforeseeable uncertainty—or unk unks—is present. In order to understand the nature of the limitations, we begin by observing that the standard PRM approach rests on a fundamental assumption—namely, that we are operating essentially on *known terrain*, where it is known, in principle, what events and outcomes of actions to expect, and with *moderate complexity*, where the nature of the "solution space" is roughly known, where an action does not cause entirely unexpected effects in different parts of the project, and where we can choose a best course of action. In other words, we know the *range* of things that can happen and their causes, even if we may not be able to predict with certainty which of the identified events will happen or to what degree of probability they are likely to occur.

Neither projects that contain significant novelty nor complex projects of long duration fulfill this assumption. Let us come back to the mountaineering metaphor from the introduction to Part I of the book. This time, you are planning an

Copyright Greg Mortensen, reproduced with permission.

expedition to an unknown mountain in the Hindu Kush in northern Afghanistan. No one has been there; there are no maps and no weather forecasts because there are no meteorological stations close by. Mastering the unknown mountain requires more sophisticated mountaineering skills, as well as more experienced and flexible people who can observe the terrain and the weather during the expedition and who can make decisions in response to what they learn. Committing to any plan, no matter how sophisticated, will lead to trouble if the team is not ready to substantially deviate from it if necessary. The team may even decide that the originally targeted mountain is too difficult in the given weather conditions and switch to an adjacent lower peak (learning). Or the team may prepare two alternative routes in parallel and keep both of them "alive" until the final approach of the summit (selectionism).

The mountaineering example is representative of projects that are novel in terms of the technology employed and/or the markets pursued, and projects of long duration. These are commonly plagued by fundamentally unforeseeable events and/or unknown interactions among different actions and project parts.[1] A "straight" application of PRM, without recognizing the additional novelty challenge, is insufficient and may have destructive effects, as we saw in Chapter 2. In Chapter 3, we have discussed the fundamental sources of uncertainty, and we have proposed an extension of PRM, including control-and-fast-response, project contracts, and, in particular, the two fundamental approaches of responding to unk unks, namely, trial-and-error learning and selectionism.

We are, of course, not the first to observe that project management and PRM must be adjusted to the presence of uncertainty and complexity. "Contingent approaches in project management" became a major theme in the second half of the 1990s. For example, in 1999, Terry Williams stated, "Projects are becoming increasingly complex; traditional project management methods are proving inadequate, and new methods of analysis and management are needed."[2] Empirical studies have shown that complexity and uncertainty lead to budget and schedule overruns and to high costs of the system developed.[3]

Miller and Lessard, in their study of large engineering projects, concluded, "The assumption that large engineering projects can be scoped, planned and managed with existing planning techniques cannot prevent problems, which are then seen as managerial failures. Prior empirical studies . . . have focused on technical and economic factors, but few suggested that the model of pre-specified rational planning is increasingly in trouble."[4]

Aaron Shenhar and his coworkers[5] identified "system scope" (and thus complexity) and uncertainty as major drivers of project management approaches: Complex projects require stringent planning and control, even bureaucracy. Highly uncertain projects require flexibility, testing, and intensive communication [we would call this "learning"]. Projects that exhibit both uncertainty and complexity need systems engineering, integration, and risk management. We build on this work in the context of PRM, by proposing operational principles and methods of management.

It is now the mission of Part II of this book to explain how these approaches to managing unforeseen uncertainty and complexity work in practice. The first question we have to answer is this: If unk unks are "unforeseeable," how can we prepare for them? Isn't that a contradiction, preparing for something that is unforeseeable? Chapter 4 addresses this question. It demonstrates the *diagnosis* of the types of uncertainty and complexity at the outset of the project. Yes, unk unks are fundamentally unforeseeable, but their *presence* can be predicted by diagnosing gaps in the team's knowledge about the project. Any important gap means that unk unks lurk (although the team can't know what they are). A systematic diagnosis of uncertainty and complexity is shown in the example of Escend, a startup company that tackled a novel market.

Chapters 5 and 6 then illustrate what trial-and-error learning and selectionism look like in real projects. Chapter 5 presents a project that used the learning approach (the startup company, Escend, from Chapter 4), and Chapter 6 presents a project with selectionism (another startup company that developed a new technology).

The question then arises: If we have two approaches to managing unk unks, how do we choose between them? Which one should be used when? Chapter 7 explains the trade-offs between them, on the cost side—How much does it cost to perform parallel selectionist trials versus experimenting and learning—and on the value side—How good might the solutions be that each approach produces? We also emphasize that in many cases, this is not an either/or question, but both approaches together provide a powerful combination of responding to unforeseeable uncertainty. Chapter 7 offers a decision framework for deciding when to use which approach and when to combine them, and we illustrate the framework on the Circored example from Chapter 2.

Endnotes

1. See, for example, Morris and Hugh 1987, Schrader et al. 1993, Hamel and Prahalad 1994, Miller and Lessard 2000, or Pich et al. 2002.

2. Williams 1999, p. 269.

3. For example, in the context of product development projects, see Tatikonda and Rosenthal 2000. For large engineering projects, see Miller and Lessard 2000. See also Morris and Hough's (1987) classic study of large projects, which also found that uncertainty and complexity cause problems.

4. Miller and Lessard 2000, p. 4.

5. See, for example, Shenhar 1998 and 2001, Shenhar and Dvir 1996, Dvir et al. 1998.

Diagnosing Complexity and Uncertainty

We saw in Chapter 2 how the presence of unforeseeable factors made traditional PRM techniques insufficient for the Circored project. We then argued in Chapter 3 that a broader view of PRM is necessary, one that allows for unk unks combined with project complexity and sets up a different project management approach, selectionism or learning, at the outset.

But the million-dollar question is this: How can unk unks be diagnosed at the outset if they are, by definition, "unforeseeable"? Our answer is that, in essence, managers must ask themselves, in all honesty: "What do I know, and where do I have fundamental knowledge gaps?" In every knowledge gap lurk potential unk unks.

In this chapter, we show how to diagnose unk unks and complexity at the beginning of the project. First, we illustrate how diligent and frank self-examination can enable project management to recognize the areas of knowledge gaps and focus their attention on them. We illustrate the diagnosis of unk unks and their subsequent flexible management with the crisis situation faced by one Silicon Valley startup company, Escend Technologies. A startup venture is a special type of project that starts with the first funding decision, follows a sequence of delivery and funding milestones (see Table 5.1 in Chapter 5), and ends when a merger, acquisition or IPO, or a transition to a self-financing and profitable ongoing concern has been achieved—or when the business closes. In Section 4.5, we turn to the diagnosis of complexity, using a systematic tool, the design structure matrix. At the end of the chapter, we link complexity back to the Escend example.

4.1 Diagnosing the Unforeseeable at Escend Technologies[1]
4.1.1 Background

Escend Technologies was founded in 1999 to enable electronic component manufacturers to connect and collaborate with manufacturers' representatives (reps), intermediaries that sold the components to electronics OEMs.[2] During the 1990s, component manufacturers, OEMs, distributors, and contract manufacturers in consumer electronics were becoming increasingly fragmented and disconnected as a result of outsourcing and globalization. There was a need to create a common customer record that tracked new products through the design, prototype, and manufacturing phases across the multiple companies involved.

By mid-2003, Escend was floundering, having burned through $16 million in venture funding, and requesting $6 million more to continue operations. Escend had won 20 rep firms as customers, who were supposed to pay—but did not—ongoing usage fees of about $800,000 per month. Sales had stagnated a year earlier, and the cash burn rate had grown to $650,000 per month. The CEO had no concrete plans, only vague promises, for any improvement. However, he reported progress on the product and the sales pipeline and maintained that it was only a matter of time before the company would be on its feet. When the initial response for the $6 million request was a resounding no, the board thought that the message, not the business or the management team, was the reason and behaved as usual—packaging the company for another financing round (i.e., improving the message by rewriting the presentation) and making introductions to their networks of contacts.[3]

Before completing the next round, the board (comprising the key investors) had to critically assess management's capability and the cash burn rate. It quickly became clear that the leadership had to change, as there was no convincing planned way out of the crisis, and business irregularities were emerging. Elaine Bailey temporarily stepped in as the CEO in mid-July 2003 to turn around Escend.[4] She was a general partner at Novus Ventures, a small business investment corporation, and one of the key investors. Within three weeks, she had to decide whether or not to recommend that Novus participate in another round of financing for Escend. A "yes" decision would give Escend a chance to fix its problems, become successful, and provide Novus with a positive return on investment. A "no" decision would end Escend's existence when the remaining cash was spent, and Novus would lose all it had invested in Escend. Writing off Escend might reduce Novus' chances of drawing on additional SBA funds for future investments. But investing in the next round and then losing everything (including other Escend investors' funds and Elaine's credibility in the VC community[5]) would be much worse.

4.1.2 Diagnosing the Unforeseeable

By examining the companies' expense reports and telephone records and by interviewing the entire team (35 employees) and four key manufacturers, Elaine quickly acquired a feel for the situation. She realized what a mess she had inherited. The CEO had invited buddies into the management team who seemed more interested in paying themselves perks than in pushing the business. The product functionality had major flaws (for example, customers could not specify custom reports), the architecture was inflexible (running in batch mode, not real time) and difficult to scale, and the Oregon-based design team was uncooperative. The phone logs revealed that the salespeople spent little time talking to customers and prospects. Elaine discovered that the $15 million pipeline was grossly inflated (the actual figure was $256,000) and that no potential customers existed. Additionally, morale had hit rock bottom.

Elaine sought the advice of her coauthors, two academics with whom she regularly exchanged ideas. "What should I do? Some of the problems and urgent moves are crystal clear. But I have no idea where the real fundamental problem lies. Perhaps it's just execution, but perhaps it's something to do with the industry. It's all opaque, like trying to see through a rock."

Her coauthors advised her against asking the question of what to do at this stage, as the situation was too uncertain and unstructured. They recommended first that she ask herself a fundamental question: "What do I know, what do I think I know, and where am I completely ignorant and unsure?" In other words, they recommended questioning all assumptions and identifying areas of the mess where she was on firm ground, as opposed to areas where she had major knowledge gaps.

This prompted Elaine to examine each part of the business and identify open issues. Some, such as reducing the burn rate and the head count, were obvious for an experienced VC partner, and the actions to address them were very clear. However, she began to realize that the fundamental concern was not within the company but in the question of, "Is this a fundable business?" Writing down the problem areas resulted, after some consultation and iteration, in the diagnosis summarized in Table 4.1.

Problem area 10 wasn't a problem at all; rather, Elaine found out that the customer support team was one of the key assets of the company. Problem areas 4 through 9 were relatively straightforward—they contained risks, but it was quite clear what had to be done. Problem area 3 was more difficult: The software product had some functionality gaps, and it was not entirely clear how to plug them. However, a hypothesis was already on the table, namely, that the rigidity might be mitigated by an interactive shell behind which the batch program would run.

Table 4.1: Problem Areas with Uncertainty Type

Problem Area	Situation	Uncertainty
1. Customer need (external)	Would customers buy Escend's products? Why? What is the customers' pain?	High potential for unk unks
2. Industry readiness (external)	No collaboration SW play has succeeded, willingness of players' openness questionable, what do other players want?	High potential for unk unks
3. Product functionality (external)	Holes in functionality, too rigid, based on once-per-day batch mode when customers wanted real-time product	Foreseeable, possible unk unks
4. Cash burn rate (internal)	$650,000 per month	Variation
5. Top management team (internal)	Complacent and dishonest, risk of lawsuit	Foreseeable
6. Sales team (internal)	Not sufficiently active; inflating results	Foreseeable
7. Head count (internal)	35 employees (70 at the high point, including 50 software developers), lack of performance	Foreseeable
8. Geographically dispersed operations (internal)	Sales offices in Alaska, California, Massachusetts, New Hampshire, and Texas; development team in Maine, Nevada, Pennsylvania, and Oregon	Foreseeable
9. Design team (external and internal)	"Old style" with outmoded (monolithic) architecture approach, uncooperative	Foreseeable
10. Support team (external and internal)	Strength of the company	Not a problem

The fundamental knowledge gaps lay in areas 1 and 2. The technology enabling Escend (XML and Rosettanet) had not existed before 2000. XML made collaboration possible from one database to another, and Escend aimed not only at supply chain management but also at demand creation across multiple countries. There were no competitors in this space; no one had defined the problem before, and no analysts were covering any companies in this part of the industry. It was uncharted territory—no one had gone there before. As a consequence, customer needs were undefined, and the ability of the players in the electronics industry to collaborate via common software was unclear. Thus, the required functionality was unknown. Customers could not articulate their needs, and different players would name mutually incompatible benefits because no one understood where the product would ultimately create the most value.[6]

In parallel to the planning subprojects, the knowledge gap around customer needs and the readiness of the industry for a player like Escend became a learning project. Elaine reserved part of her and the Escend team's time to reflect and to gather information from multiple parties about that problem area, not knowing what to expect. She remained open to finding nothing of significance in this inquiry or something that might prompt her to fundamentally rethink the business model—or even to shut Escend down.

During her first three months as CEO, Elaine traveled frequently to interview enterprise firms, end customers, analysts, consultants, VCs who did not invest (through not believing in either the management or the business model), managers of various collaboration startup companies, and academics. She searched and probed to find out why collaboration solutions had not succeeded, what the needs of the enterprise customers were, and what problems the complex industry structure really posed. In face-to-face meetings, people opened up and provided useful nonverbal cues—information that could never be obtained by only attending board of directors' meetings. Slowly, information emerged, but the information kept changing as the industry evolved.

After an intense three-week probing of the problem areas where unk unks threatened, Elaine made a leap of faith to conclude that: (a) demand for Escend's product existed; (b) the potential market was large enough to have the potential of a sufficient return on investment; (c) the competitors were far enough behind for the opportunity not yet to have been tapped; and (d) Escend represented one of the last remaining large enterprise software "plays" for VC investors. Elaine recommended that Novus provide additional funds, and in August 2003, Escend closed a $7 million round from Novus and NIF Ventures, an affiliate of Daiwa Securities.[7]

Thus, Escend was allowed to continue. According to the problem characterization in Table 4.1, Elaine Bailey managed part of it as planning projects, and parts of it as a learning project, in which she did not know what she would find. The management of the learning project is described in Chapter 5. In the remainder of this chapter, we generalize from Escend's initial turnaround phase and suggest how a project management team can go about diagnosing unk unks, as well as complexity.

4.2 Diagnosing Complexity

In addition to uncertainty, we argued in Section 3.4 that complexity is an important characteristic of projects. Complexity captures the number of interactions among different project pieces, and high complexity hinders the team's ability to "optimize" the overall solution. Thus, the project team needs to diagnose complexity at the outset, just as it diagnoses the uncertainty.

A practical tool, in the form of the *design structure matrix* (DSM), exists to map and illustrate the complexity of a project.[8] The DSM maps interactions among pieces of a project in matrix form, by listing which task needs input from which. Interactions may take the form of, for example, *information* (one task requires input from the other), *space or resources* (both tasks take place in the same spot or use the same resource), *material* (one task provides material that the other uses), or *constraints* (one task may produce a system parameter that constrains the things that are allowed in the other task).

Figure 4.1 presents a brief example of the application of this tool. In the figure, *impacted* tasks are listed in the columns, and *impacting* tasks along the rows. Crosses (x) mark dependencies. Task B is *sequentially dependent* on task A, as the impact goes only one way; for example, task A produces information that task B needs, or task A occupies space that task B must then avoid. Tasks B and C are *independent*. Tasks C and D impact each other—that is, they might require mutual information input. Thus, they are *coupled* (interdependent). The matrix suggests an ordering of the tasks: A, then B in parallel with C and D, the latter two being performed in a closely coordinated way.

Interactions occur in three domains simultaneously: the *system domain*, in which components interact, the *task domain*, in which project activities interact, and the *organizational domain*, in which teams and stakeholders interact. For each of the three domains, a separate DSM interaction matrix may be drawn up. All interactions are relevant for the project and may contribute to overall complexity, and yet it is useful to distinguish the domains because they are concerned with different types of interactions. Table 4.2 lists examples of interactions for the three domains.

	A	B	C	D	
Task A	A				A-B: Sequential
Task B	x	B			B-C: Independent
Task C			C	x	C-D: Coupled
Task D	x		x	D	A-D: Sequential

Figure 4.1   Example of a design structure matrix, DSM

Obviously, the three domains are not independent; they strongly overlap and influence one another. Linking the system and task domains, many tasks have a one-to-one relationship with system components (a task comprises work on a system component), and other tasks relate to interactions among system components (for example, coordination or integration tasks). Relating the task and organization domains, tasks are typically assigned to individuals, and highly interdependent task groups are arranged in organizational units. As a result of these relationships, the DSM interaction matrices for the three domains often look related, as Table 4.2 shows.

Figure 4.2 uses three DSMs to describe the interactions in a climate control system development project.[9] The first DSM lists the components (16 major subsystems in this case) and their interactions; the second lists the task interactions (there are many more tasks than components) and the interactions among the organizational groups, the subteams. Some subteams were assigned to component groups of the system, with some overlap across teams (members belonging to multiple teams) in order to ensure integration and representation of all expertise. One team worked on the integration of the components; therefore, it interacted with all the other teams simultaneously.

Table 4.2: Examples of Interaction Types in Three Domains

System Domain (among system components)	Task Domain (among project tasks)	Organizational Domain (among actors or groups)
Spatial (e.g., components occupy the same space)	Spatial (e.g., two tasks work on the same system component and cannot, therefore, proceed simultaneously)	Groups are affected by the task domain interactions because they must carry out the tasks
Energy (e.g., one component supplies the other with electricity)		
Flows (e.g., one component transfers fluids or other materials to the other)	Resources (e.g., two tasks compete for access to the same test facility, or need the same expert personnel)	In addition: Power, status, influence (e.g., the group that "wins" in a resource competition has more status)
Force (e.g., one component exerts pressure on the other)	Artifacts (e.g., a task delivers a prototype for further development)	
Information (e.g., in software, or control units)	Information (e.g., one task delivers information that serves as input for the other task)	Goals (e.g., groups' interests are incompatible: one group wants higher performance; the other wants lower costs)
Synchronization (e.g., two components must deliver a flow of signals simultaneously)		

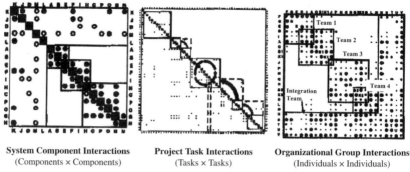

System Component Interactions	Project Task Interactions	Organizational Group Interactions
(Components × Components)	(Tasks × Tasks)	(Individuals × Individuals)

Figure 4.2   Interactions in three domains in a system development project

It is not surprising that the three DSMs in Figure 4.2 are related—component interactions drive task interactions, which, in turn, are mirrored in team interactions.[10] However, the similarity is not perfect; some task interdependencies are introduced by the way the project is organized, and the project tasks ignore some component interactions (for example, those that are judged less critical). Similarly, interest conflicts among teams may be independent of the system that is being built.

The DSM becomes heavy and cumbersome when it is used in a large project to assign tasks and personnel in detail: When the matrix grows beyond a size of 50 x 50, it becomes difficult to administer. However, the DSM is an excellent tool for diagnosing complexity at the outset of the project, at a semiaggregated level:

▲ Identify the main subsystems (or components) of the system that the project builds (no more than about 20).

▲ Use the work breakdown structure to identify major (aggregate) tasks of the project (no more than about 40).

▲ Use a stakeholder mapping to identify the major project subteams (no more than about 5) and the other stakeholders (no more than about 15).

▲ Construct the three DSMs, even if the interactions have to be estimated.

Complexity can now be roughly estimated as a product:

Complexity = (Sum of all project elements) × (Sum of all interactions)

Here, the sum of all project elements includes all three domains, subsystems, tasks, *and* stakeholders; similarly, the interactions include all three domains. First, this reflects the fact that complexity can arise from different sides, the system architecture, project organization, and the stakeholder interest constellation. Second, this complexity measure reflects the fact that complexity is driven by the combination of the number of elements and their interactions. In other words, if there are very few interactions, complexity is low: Even a large project (with many elements) is not complex if the elements do not interact; the elements can be tackled one by one. If the project is very small (few elements), complexity is low, even if those elements heavily interact. On the other hand, the number of elements and their interactions compound one another; adding a few interactions to a large system makes complexity *much* worse.

This measure of complexity refines the distinction of "task complexity" and "relationship complexity" that we made in Chapter 3. Moreover, this measure can be estimated (albeit roughly) at the outset of a project, when a first work breakdown structure and project organization have been determined. Complexity heavily influences the solution approaches that are feasible in a project, and, moreover, complexity and uncertainty interact in their impact on the best project approach, as we will discuss in detail in Chapter 7.

4.3 A Process for Diagnosing Uncertainty and Complexity at the Outset

In Sections 4.1 and 4.2, we showed how unk unks and complexity can be identified at the outset of a project. We now attempt to make this identification systematic in a *diagnosis process*. Identifying complexity requires listing the project elements and estimating their interactions. Identifying unk unks poses the apparent paradox of identifying something that is, by definition, unforeseeable. In essence, resolving this paradox requires asking: *What do I know and what do I not know?* and *Where are the major knowledge gaps?* Every knowledge gap poses the potential of unk unks. The diagnosis process is outlined in Figure 4.3.

1. First, identify the problem structure, as Elaine Bailey did when she identified the mess she had inherited at Escend: Do we understand the ultimate goal for the project as it is currently defined? Who are the stakeholders that may influence the outcome of the project? Do we have some understanding of the causality of actions and effects in the project?

2. Then, break the overall problem ("How do we generate enough sales?") into pieces: What are the modules or subprojects that require our attention? Who are the major players and/or stakeholders? At Escend, Elaine broke the problem up according to the market forces (customers, competitors, and partners) and the functional teams in the organization (sales, cash management, design, and so on; see Table 4.1.

3. For each piece, perform risk identification. In other words, assess the scope of variation in budgets, schedules, and achieved performance, and identify major risks that need managing (with standard PRM methods). Also, identify knowledge gaps, by probing assumptions and asking what you know and what you do not know. This identifies areas of potential unk unks. As the events at Escend demonstrate, this is a highly iterative and gradual process.

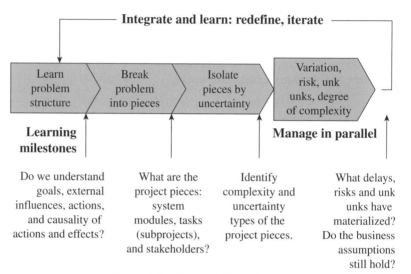

Figure 4.3   A project diagnosis process

4. Estimate the complexity of each project piece, and of the overall project, from the respective number of elements and interactions. At Escend, complexity was not a major consideration; the interactions between the system configuration and the industry players were understandable. The challenge stemmed from the unforeseeable uncertainty of not knowing where this market was going.

5. Finally, manage the pieces in parallel, using different management approaches according to the combination of complexity and uncertainty: planned execution and traditional PRM for the pieces with foreseeable uncertainty, some selectionism for the highly complex subprojects, and a combination of selectionism and flexible iteration for the pieces that are threatened by a combination of complexity and unk unks. We have not yet discussed what the "right" combination is; this will be the subject of Chapter 7. As the startup progresses toward the next milestone, or IPO, new information emerges, and the problem pieces and approaches must be updated and modified.

We have found it helpful to summarize the characterization of the project pieces (the third learning milestone in Figure 4.3) with *uncertainty and complexity profiles*. Project managers and their teams do not need to have an exact analysis of the uncertainty and complexity challenges they face to decide upon the appropriate management tools to use. What they need is a prioritization tool that helps them to define the relative importance of each source of influence at the starting point of the project, and eventually to evaluate how the portfolio of complexity and uncertainty influences is evolving over time.

After identifying the uncertainty and complexity types in Step 3 of Figure 4.3, we score the importance of each influence on a scale of 1 to 10. The importance can be a potential impact or the amount of time that the team thinks they need to spend on understanding and addressing them. Figure 4.4 demonstrates the idea with three brief examples of uncertainty and complexity profiles, summarized at the level of entire projects.

The project on the left represents the construction of a cruise ship at a major French shipyard. In naval construction, many contractors perform tasks in parallel; it is easy to imagine a site with 2,000 people from 500 subcontractors working at the same time building the restaurant, the casino, the cabins, the swimming pool; installing the wiring for the Internet system; tuning the engine; and so on.

The components of the ship do not strongly interact, and many of them can be built separately, off-site. The key interactions, for components as well as tasks, are space constraints and supply connections. Most of these interactions can be managed by defining good interfaces. A large project size, combined with relatively few interactions, implies medium to high (not extremely high) complexity. Thus, a delay or quality problem of one contractor may cause a chain reaction of problems for other contractors. This level of complexity already requires diligent coordination among many subcontractors who work on the site at the same time.

Variation due to bad weather, or quality problems or delivery delays by suppliers, may influence the execution of the plan. Variation is therefore estimated as medium. This requires buffer management to control schedule variation. However, for most of these cruise ships, the design is well known. There are really no significant gaps in knowledge, and therefore, the potential for foreseeable and unforeseeable uncertainty is limited.

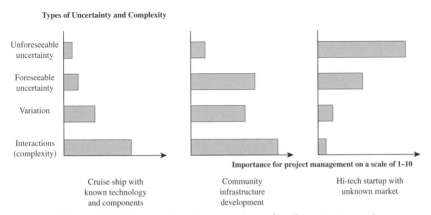

Figure 4.4 Uncertainty and complexity profiles: Prioritization tool

The project in the center of Figure 4.4 refers to a residential community construction project, Ladera Ranch in Southern California. This was a multiyear, several-hundred-million-dollar earth-moving infrastructure (water, sewers, electricity) development project to prepare the construction of a new community of several hundred houses. The project team moved millions of cubic yards of earth to provide independent builders with house pads, streets, water runoffs, landscaping, and utilities.[11] Their major objective was to plan the cuts and fills in a way that moved earth the shortest distances possible.

Since the project could be completed before other builders moved in, the team had much freedom to work at multiple places in parallel; in other words, task complexity was low, and thus, coordination was easy to ensure. However, the complexity of the various "components" (the individual lots) was very high: The efficiency of operations in such an earth-moving project was influenced by soil conditions, in particular, moisture and composition. Moist soil required more excavation and took longer to settle before one could build on it, so the project team had to consider mechanically drying the dirt rather than delay selling lots. Some types of soil required different slopes for stability, influencing the available amount of flat area for houses and streets.

This could all be planned, in principle, by drawing up a PRM contingency plan for each scenario of the type: "If the soil is moist and type x at location y, do plan A; if it is dry and soil type z, do plan B." But in practice, this contingency planning was infeasible because of the interdependent nature of cuts and fills across locations. The number of scenarios proliferated with the number of locations considered, making the moisture level and exact soil type, in effect, unpredictable. Thus, the Ladera Ranch team was forced to dig and then, based on their experience, react to what they found. In effect, component complexity forced an experimental approach. Thus, complexity and foreseeable uncertainty were closely related, and both rated high in this project.

However, truly unforeseen events, which would completely alter project execution, were rare. Examples of such events would be the discovery of prehistoric Indian ruins or a rare animal or plant species. Therefore, the score for the relative importance of unforeseeable influences, shown in Figure 4.4, is low.

The project in the right-hand panel of Figure 4.4 refers to an Internet startup company that was created in 1999 in order to apply the Priceline reverse-auction business model to the German market. A major challenge for this company was the unpredictable reaction of German consumers to this business model.[12] Despite numerous changes in the selling process to accommodate the specific preferences of the German consumer, by mid-2000, the company could see that the consumer auction boom was faltering.

Knowing that it could not survive on customer-driven pricing alone, the startup radically changed its business model. It developed software services for industrial customers and an Internet-based ticket search engine for travel agents. By summer 2001, this search engine, which dynamically optimized offers from multiple airline reservation systems, had become the most promising of the company's offerings. The change in the business model was quite dramatic. In response to the redefinition of its mission, one of its investors commented, "How can they change the business model this much? It is like we gave them money to develop a sausage factory, and now they tell us they have moved into building fighter planes." In the end, the company succumbed to the decline of the travel business after the burst of the Internet bubble and gave half of the invested money back to the investors in 2002.

In this startup company, neither complexity and coordination nor variation, foreseeable uncertainty, or managing the daily operations was the biggest challenge. The biggest challenge was to keep the finger on the consumers' pulse, to see how they perceived the business model and how their tastes evolved. The challenge was to cope with the unk unks in a new market.

Our approach, as summarized in Figures 4-3 and 4-4, is related to *discovery driven planning*,[13] which proposes to explicitly acknowledge unk unks and to uncover them with four analyses: (a) a reverse income statement calculates what would have to be achieved in terms of market share and revenues in order to reach a given return target; (b) a pro forma operations specification shows the key steps for producing the desired output and asks whether these steps can be performed with "normal" process capabilities (or whether heroic feats are required); (c) an assumptions checklist compares the plan with experiences in similar situations or with expert advice (e.g., "we assume the average selling price to be around $1.60—is that justified?"); and (d) milestone planning anticipates at what points which risks can be eliminated.

Discovery driven planning is very useful, but has two limitations. Examining the estimated causal structure of profits is an obvious start but runs the risk of missing the most dangerous unk unks: The venture's plans are carefully worked out, everything is consistent and plausible, and still, unexpected factors emerge later that make the entire business model obsolete. Moreover, "checking assumptions" means that we ask for each parameter whether it is really fixed, or whether we can diagnose a relevant unk unk (an influence that may change and influence the outcome). Again, this is a good start, but may lead to the "double blindness" that we have discussed in Chapter 3, and it will not suffice when causal relationships are unknown and influences may arise that the team cannot identify through critical questioning at the outset. Thus, discovery driven planning demands a level of knowledge and structure that may not be available. It does not utilize knowledge and intuition possessed by the management team, which may be less precise than identifying individual influences, and which we propose to tap.

4.4 Evolve the Complexity and Uncertainty Profile

For some unk unks, identifying their *possibility* immediately identifies *them*, and thus shifts them to identified risks. For example, realizing that patients may not comply with the restrictions of use for a newly developed drug immediately transforms this from an unk unk, something the team did not consider, to an identified risk. This is, of course, precisely the purpose of the risk identification stage of established PRM methods. However, in novel projects, there are, typically, areas of knowledge gaps that can be identified as such, without being able to identify the risks themselves, as we have seen in the Escend example and in Section 4.3.

Let us consider another example. Many companies that entered the Chinese market in the late 1980s or early 1990s knew that they had little knowledge about doing business in China, and that most consultants who claimed to be able to help were unreliable. In other words, these companies were aware of their knowledge gaps and of the presence of unk unks. Some companies tried to protect themselves against the unk unks by teaming up with a "reliable" local partner. Disturbances were still to be expected in the implementation of the "project" of market entry. But they were expected to be reduced to variance or foreseeable influences, because the Chinese partner was brought on board with sufficient experience to handle these disturbances. To the dismay of many early entrants, these partners themselves often turned out to be sources of unforeseeable uncertainty: They knew far less about the market than they had pretended, they were not prepared for sudden changes in the policies imposed by the provincial or municipal governments, or they had a hidden agenda that made them unpredictable.

A key rule in using complexity and uncertainty profiles is that, over the course of the project, they change as the team learns. At critical transition points in the project, major uncertainties emerge or are resolved, and only after significant progress is made will the team know what is ahead. This is summarized in Figure 4.5.

Let us take a startup company, such as Escend, as an example. It will go through various stages of its existence. It may start as a simple idea in a university laboratory. A business angel may be prepared to put in some time and money to develop the initial concept. Once the concept proves attractive, the real startup happens. The product is developed but may not be completely operational. Customers are tested; if their reactions are positive, the company can obtain additional financing. The product is launched and tested, and the first revenues come in. Once this stage is successfully passed, the company expands and may become ready for an IPO.

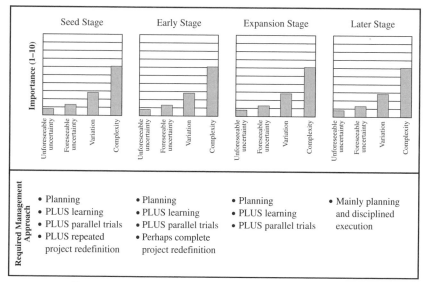

Figure 4.5 Evolution of complexity and uncertainty profile

The types of uncertainty change as the company progresses. In the first stage, when the idea is still the dream of a few founders, complexity is almost nonexistent, but unk unks are everywhere: How will we find the technical solutions? What shape could the market take? How do we best finance the project? And so on.

Bringing in a business angel with a lot of experience reduces some of the unk unks to foreseeable uncertainty. For example, a business angel may be able to help reduce the market uncertainty and propose contingencies in the financial construction of the company (like a decision tree, as illustrated in Section 3.3): "If we can obtain this financing, we may also be able to convince that party." For products that push the technological boundary, unk unks about technology and market definition may still exist. However, the relative importance of the various sources of uncertainty is evolving.

Once first customer tests have succeeded, market uncertainty may well be redefined as variance: The direction of the market is becoming clear, but it is still uncertain at what speed customers start adopting the product. Once the startup enters the preparation stage of an IPO, management attention shifts from managing unk unks and foreseeable uncertainty to managing complex relations with investors, investment analysts, partners, and the stock exchange, not to mention the lawyers.

4.5 Conclusion

In this chapter, we have extended the risk identification phase of PRM to include the diagnosis of complexity and the identification of knowledge gaps, signifying the presence of unk unks. In the example of Escend, we have shown that, although unk unks themselves are unforeseeable, the project management team can recognize knowledge gaps. The areas of knowledge can then be searched and probed, in parallel with project execution.

The project profile diagnosis, complexity identification, and probing for unk unks can be pursued as a systematic diagnosis process. This process is essentially a search with an unknown outcome, so it is inherently iterative. The result of the probing can be summarized via a simple qualitative representation tool, the complexity and uncertainty profile, to estimate the relative importance of each type of uncertainty. The uncertainty profile allows the project management team to choose the best management approach for the project.

Endnotes

1. This section is based on Loch, Solt, and Bailey 2005.

2. OEM stands for "original equipment manufacturer." OEMs sell products to the end consumer. OEMs often perform marketing and system design but outsource component design and manufacturing.

3. The investors, accepting top management's view, thought maybe that the Escend story was just not being told well, and they, led by Elaine Bailey, rewrote the executive summary, generated a "VC presentation," and taught the CEO how to give the presentation. The result was zero interest by potential new VC investors.

4. Elaine's experience running her own rep firm in the 1980s made her the logical choice and meant that she knew the "right" questions to ask—or that she would discover what they were.

5. The lead VC is primarily responsible for performing the due diligence that supports a recommendation, and in the tight-knit VC community, bad decisions (such as "throwing good money after bad") tarnish reputations.

6. It is well documented that technologies that enable new uses in the market cause unforeseeable effects, see, e.g., O'Connor, G. C. and R. Veryzer 2001; see also Leonard-Barton 1995.

7. Elaine reports that she almost threw in the towel in the run-up to the recommendation. The situation was too confusing and opaque. The discussion of unk unks and flexible adjustment to what lay ahead gave her the mind-set of possibilities, and the will to continue.

8. The DSM was first proposed in engineering by Steward 1981. Eppinger et al. 1994 developed it further as a management tool. The illustrative example in Figure 4.1 is based on Loch and Terwiesch 2000.

9. Figure 4.2 is adapted from Eppinger and Salminen 2001.

10. Sosa et al. 2005 quantify the similarity among the three matrices in a statistical sense and find that there is, indeed, substantial overlap. Moreover, these authors show some first evidence that congruence between system, project, and organizational architectures lead to higher performance of the project team. The work by these researchers has emphasized the usefulness of the DSM for assigning tasks and teams in a way that minimizes interactions, and thus complexity. In this chapter, we emphasize the usefulness of the DSM as a diagnosis tool (at a less precise, more aggregate level) at the outset of a project.

11. For further details, see De Meyer et al. 2002.

12. For example, Germans did not want to pay via credit card (habituation took four years), and many did not want to commit to an offer without knowing the price. Other challenges were the fact that business processes could not be patented in Europe, and the relative scarcity of venture capital in Germany. The company invited "strategic investors," large companies who had an interest in developing alternative Internet sales channels, in the hope that they would support the startup's growth. This turned out to be a misjudgment, as it dragged the startup into some large-company internal resource struggles that prevented effective support by the strategic partners. For details, see De Meyer et al. 2002.

13. Discovery driven planning was developed by McGrath 1995, and McGrath and MacMillan 2000.

Learning in Projects

Much has been written in the academic and popular business press about learning in organizations, and learning has been given many definitions.[1] For our purpose, we use a very specific definition of learning, one that emphasizes adjustment in response to emerging unk unks in a project:

> Learning in projects is the flexible adjustment of the project approach to the changing environment as it occurs; these adjustments are based on new information obtained during the project and on developing new—that is, not previously planned—solutions during the course of the project.

The essence of this approach to project management is that each new activity will provide new insights and information, which can be used to review and revise the project plan, the resources required, and the stakeholders to be dealt with. While each of the changes may be minor, the project itself may look quite different at the end from the original plan and intention. This type of learning "as we go" should not be confounded with projects as instances of learning about the organization (because you do something out of the normal activities of your organization) or post-project audits that are used to improve future projects.

To understand learning "as you go," let us briefly return to the simple mountaineering example that we used for illustration in the introductions to Parts I and II. There we contrasted two mountaineering expeditions, one up a known mountain for which we have a map and a weather forecast and another up an unknown mountain with unknown weather conditions. The essence of the comparison is between executing an existing plan versus developing the plan as one goes along and learns about the terrain en route. Mastering the unknown mountain requires more sophisticated mountaineering skills, and more experienced and flexible people who can observe the terrain and keep an eye on the weather conditions during the expedition and make decisions in response to what they learn. More information needs to be gathered, and coordination among all stakeholders must be more flexible.

To demonstrate the project learning approach to unk unks, we continue the Escend Technologies example from Chapter 4.[2] There we showed how the presence of unk unks was diagnosed, and now we show how learning was managed in this project of rescuing this startup company and shepherding it to the next financing round. After this example, we discuss in further detail the process of project learning.

5.1 Learning at Escend Technologies

When Escend Technologies was founded in 1999, it was like most other startups in that it had a business plan. The business plan was an essential part of selling the Escend business proposition to potential investors, as is common practice in new venture funding (see Table 5.1 for a view of how venture capitalists expect a new venture to progress through various development stages to the initial public offering, or IPO). The business plan is not unlike a project plan in that it contains, in addition to a description of the market and core product, a development plan consisting of key milestones to be met by the startup at key points in time.

Monthly board of directors' meetings track progress according to the key milestones. The board aims at providing guidance and continuity in governance but often considers only what has occurred since the previous meeting. The board hears management's version of the world—that is, problems, progress, development, cash requirements versus burn rates, and so on—and if it sounds logical, the board may tolerate delays in achieving the milestones. If not, actions may be taken to correct the situation, assuming nonexecutive board members have the power to do so.

5.1.1 Planning and Firefighting

Recall that Escend was built on the idea to help semiconductor and electronic component manufacturers connecting and collaborating with their extended sales force, namely the manufacturers' representative that sold their components to electronics OEMs. The founders originally conceptualized

Table 5.1: Project Stages of a New Venture Project Reaching IPO

Criteria	Seed	Startup	Early Stage	Expansion Stage	Later Stage
Definition	Research and develop initial concept	Concept or product under development, but not fully operational; usually in existence less than 18 months.	Testing or pilot production; may be commercially available and generating revenues; usually in existence less than three years	Production and commercial availability with significant revenue growth but may be without profits; usually in existence more than three years	Wide availability with ongoing revenue; likely profitable with positive cash flow; includes spin-outs of operations of existing private companies
Milestones	Business plan	Full team; prototype	Beta test; booking orders	Customers	Profitability
Major source of financing	Founders, family, and friends; angels	Angels and venture capital	Venture capital	Venture capital	Venture capital and buy-out funds
Financing required	R&D	Product development	Capital assets	Capital assets and working capital	Capital assets and working capital; product development
Operating cash flow	Negative	Increasing cash burn rate	Increasing cash burn rate	Decreasing cash burn rate	Positive
Risk	Higher				Lower
Valuation	Lower				Higher

Source: PricewaterhouseCoopers MoneyTree

the opportunity as one of collaboration among industry players who would want to be part of Escend's B2B (business-to-business) community. Figure 5.1 illustrates Escend's initial understanding of the market opportunity. In retrospect, two things become painfully obvious from this figure: First, the web of relationships among the various industry players is complex, and second, Escend had no clear idea of its value proposition to its potential customers.

In a way, it is not surprising that by mid-2003, Escend was on the brink of bankruptcy. But how did the company get into such a state? One could discuss specifics, but the general answer is that the planning approach, as laid out in Table 5.1, was not suitable for a startup like Escend. Escend faced too many unk unks: in its technology, in the industry, and in the customer needs. The milestones laid out in its business plan were simply unrealistic, not in the sense of too much stretch, but simply by virtue of the fact that no one had done this before and so no one knew what to expect. Faced with time pressure and market dynamics that they did not understand, the team was forced to improvise around the plan, and they "[found] themselves frustrated by the simultaneous pressure to act and the inability to understand what [was] going on around them."[3]

This is not unusual for startup venture projects. Drucker observed already in 1985, "When a new venture does succeed, more often than not it is in a market other than the one it was originally intended to serve, with products and services not quite those with which it had set out, brought in large part by customers it did not even think of when it started, and used for a host of purposes besides the ones for which the products were first designed."[4]

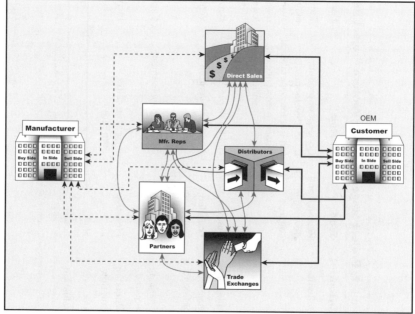

Figure 5.1 Escend Technologies' initial view of the extended sales organization

February 2000	Escend builds B2B communities that connect manufacturers with their outsourced sales channels and distributors through its sales information network.
May 2001	Escend provides the only system that overcomes the competitive disadvantages of a many-to-many business environment by speeding communication, normalizing data interchange, and connecting an entire industry.
September 2001	Escend provides the only online CRM application that includes the infrastructure and data exchange translator required for independent companies to collaborate on any aspect of the order life cycle.

Figure 5.2   The evolution of Escend's business model before 2003

During this time, Escend's value proposition went through various evolutions (see Figure 5.2). These evolutions in its value proposition were not the result of systematic investigations into the industry or the needs of its customers. They were simply reactions to events that occurred around them. They lost sight of the original objective of making money in a market opportunity and instead focused on trying to implement the business plan. The business plan became the objective, and the message was changed from time to time to help get the business plan back on track.

In fact, when the initial reaction of outside investors to the $6 million request for additional funds was negative, the board thought that the message, not the business or management team, was the problem. The board continued to behave as usual by rewriting the message to investors, without critically examining the management team or the approach to the business opportunity.

It was not until the funding situation reached a critical point that the board woke up to the situation at hand and things began to change. Unable to raise outside money, the board, including the key investors, had to assess critically both the management team and the business opportunity. The task fell to Elaine Bailey, a general partner at Novus Venture, one of the original funding partners. She faced a difficult decision: Should Novus participate in another round of funding for Escend—thus risking throwing good money after bad—or should they simply pull the plug and cut the losses already incurred in funding Escend to this point?

5.1.2 Diagnosis and Learning

In July 2003, Elaine Bailey stepped in as interim CEO of Escend Technologies to assess the company and recommend the next steps. It became clear to her relatively early on that major changes had to be made in order for the company to survive. The trouble was that it was far from obvious what these changes were or how they could be identified. As she put it, it was like "trying to see through a rock."

Fortunately for Elaine, and for Escend, she was sent in to assess the situation for purposes of generating a yes-or-no answer as to the additional funding. This perspective led her to diagnose the current situation, something the previous management team should have been doing all along. Chapter 4 already described in detail her steps to diagnose the situation. Her key insight was not to attempt to diagnose what needed to be done to implement the business plan, but to remind herself of the objective (to make money in a particular market opportunity) and to diagnose what they knew and what they did not know about the market opportunity and Escend's ability to take advantage of this market opportunity.

She recognized that there were significant knowledge gaps and, thus, major unk unks that had to be dealt with (see Table 4.1 in Chapter 4). The three main areas with the highest potential for unk unks were (1) customer needs, (2) industry readiness, and (3) product functionality. Customer needs had the greatest knowledge gaps because customers themselves could not articulate their needs. No one had yet understood where the "product" would ultimately create the most value. Thus, the required product functionality was unknown and the underlying technology (XML and RosettaNet) was fairly new. Finally, as there were no competitors yet in this space, no one had defined the problem before, and no analysts were covering any companies in this part of the industry. In other words, Escend was a pioneer in uncharted territory.

The three unk unk problem areas required a different mind-set at Escend than had been previously established. Elaine had to determine whether Escend made sense in the first place, what assumptions potential success depended on, and what different angles might offer new opportunities. This involved an open-ended search with an unknown result, requiring "switching gears" as compared to the execution mode that characterized the other problems (and most of what VCs usually do).[5]

Escend was creating a new market niche "where no one had gone before." Thus, unk unks had to be lurking in the unmapped terrain. The goal had to be to turn unk unks into known unknowns (foreseeable uncertainty). This cannot be done in a classic, straightforward "analysis"; it is a process of discovery over time. Table 5.2 shows some of the questions that Elaine and her team used to investigate assumptions Escend was making and to kick-start the discovery process.

The table provides a column for assumptions about the relevance of channel collaboration software for the business of the customer (that is, the electronics component manufacturer). Escend's value proposition is based on this assumption and is where Elaine and the team initiated the process by formulating probing questions. Two example assumptions and a few probing questions are listed in the table. The full list covered a large whiteboard, which was maintained in a meeting room. The management team would meet daily and weekly for one to two months to nail down the unk unks.

As Elaine traveled extensively during her first three months as CEO, interviewing enterprise firms, end customers, analysts, consultants, VCs who

Table 5.2: Probing Questions to Query Assumptions

Assumptions	Escend's Value Proposition for Its Customer (Component Manufacturer)	Probing Questions
Product is not a commodity/not custom silicon.	Keep customer from losing orders (*Customer need statement:* "*I'm losing orders from design win to production. I want my channel to give me a competitive advantage.*")	Once you have a design win, what's the likelihood that you'll get the order?
		How much do you lose (leave on the table) for design wins that don't translate into orders?
		What influence does your channel play in securing the order once it is designed in?
		What impact do you have in moving it from design win to order? (scale 1–10; least–most)
		What needs to happen to improve your design win to order conversion rate? (. . . and so on)
Channel is very diverse in skills, characteristics of firms	Provides in-depth visibility about design-win process (*Customer need statement:* "*don't understand the breadth and depth of the relationship with my customers.*")	On a scale of 1–10, how much visibility do you have of your customer base?
		What types of information are important to know about your customers?
		On a scale of 1–10, how well do you "service" your customers?
		What type of service (questions) are your customers requiring/demanding?
		What information is the customer requesting that is out of the control of the sales department?

did not invest (either because they did not believe in the management or the business model), managers of different collaboration startup companies, and academics, information slowly emerged, but the information kept changing as the industry evolved. As we described in Chapter 4, Elaine concluded that it was worth continuing and achieved a financing round in August 2003.

In October 2003, Elaine convened a "no good news" board meeting. She had realized that truly learning and using the unk unk concept was not easy, and letting the unk unks assume shape takes time. The table with assumptions and questions was erased and rewritten again and again, as new information came in. As Elaine put it, "We kept putting our ear to the ground, and we heard nothing. Slowly, we began to hear some faint hoof beats; then they became louder and louder."

In parallel to the planning subprojects, the knowledge gap around customer needs and the readiness of the industry for a player like Escend became a learning project. Elaine reserved part of her and the Escend team's time to reflect and to gather information from multiple parties about that problem area, not knowing what to expect. She remained open to finding nothing of significance in this inquiry or something that might prompt her to fundamentally rethink the business model—or even to shut Escend down.

5.1.3 First Results and Further Adjustments

In developing the business over the 12 months, two large business changes emerged, both of which changed Escend's strategy and neither of which could have been anticipated. First, Elaine learned how fast the electronic components market was becoming global. The "technical vertical" (industry-speak for the end-user industry application) had to be a complex product that was global in nature.[6] This implied a global platform and design-win tracking for component manufacturers when they were bidding for their components' incorporation into end products. A global platform, another product redesign, would consume precious funds and resources. Thinking that potential competitors would also have to redesign their product, which would put Escend in the lead, Elaine decided to bite the bullet. For instance, the redesign incorporated multiple languages and currencies, and multiple access points per customer, into the product. This also meant that Escend's target customers and growth strategy changed.

Second, any firm in the industry network (Figure 5.1) had limited visibility of the entire network. Component manufacturers, through their rep firms, could not track either sales or the process and, thus, could not ensure full and timely payment. OEM buyers cared about the design win but not about tracking it, because they wanted to buy at the lowest possible price and had to give extra profit margin to the contract manufacturers and design-win firms. OEM buyers often gave the job to *different* manufacturers who offered lower costs *after the design was "won."* As a result, manufacturers were changing the way they sold products, shifting away from reps toward relying on distributors and in effect taking back some of the activities that reps were offering. Therefore, Escend would have to build distribution functionality into the product. To this end, in October 2003, Escend produced a prototype that offered shipping and debiting, samples management, and pricing and quoting functionality. Coding would take another 12 months; the plan was to go live in January 2005.

In late October 2003, Elaine's search for information had led her to another startup (run by a competitor VC) that had a collaboration software product covering the demand cycle of the industry. The two had the potential to form a perfect match if their products could be made to work together. Elaine was confident that the two software products could be made interoperable (this was a problem area with foreseeable uncertainty). But the merger fell through because the investors of the other startup got some of their money back and wanted out.

The flexible way of proceeding, including repeated unplanned product changes and three major strategy changes (counting the aborted merger), was very stressful, and possible only because Elaine, combining the roles of chairperson, CEO, and partner of one of the major investors, was leading it. However, while it was Elaine, who had authority, access to the investors, and prior operational experience, it was the new Escend management team that implemented the process. Elaine assembled a new team in the late summer of 2003 but had to make further changes, including replacing the VP of sales again in June 2004. Over time, the team became fully engaged in the learning process. "Unk unks" is now a well-known term in the Escend offices, reflecting the new mind-set Elaine has instilled in the company.

Escend's business model slowly crystallized. Figure 5.3 is taken from a white paper that Escend published in the spring of 2004 and has a clarity that stands in contrast to the complex Figure 5.1. Tracing the flow of the design-win process is easy in Figure 5.3, and the demand cycle and demand fulfillment (i.e., supply) cycle is also clearly delineated. Unk unks no longer dominate; the industry structure now looks understandable, and the effect of actions taken can be traced. Foreseen uncertainty has replaced unforeseen uncertainty.

In the fall of 2004, it seemed that Escend had turned the corner and was becoming an excellent bet for the investors. At the end of October, the company obtained another $3 million from the existing investors, Novus and NIF Ventures. Daiwa Securities now intended to translate the product into Japanese and to go into Japan in December 2004, having targeted 12 Japanese companies. In November of that year, the plan foresaw cash flow breakeven occurring in Q2 2005. A new investor had signed a letter of agreement (term sheet) to invest another $1.5 million in Escend. And Elaine, signaling her confidence, was getting ready to replace herself as CEO.

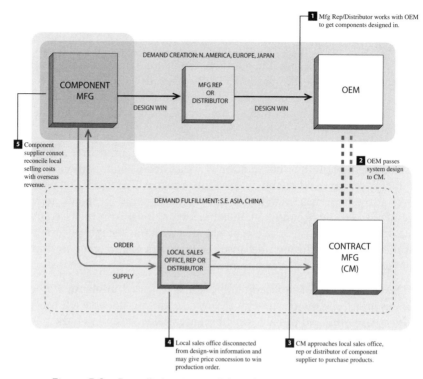

Figure 5.3 Escend's description of the industry network, spring 2004

5.2 Types of Project Learning
5.2.1 Three Types of Learning

Classic typologies of organizational learning[7] distinguish between three levels of learning: (1) single loop, (2) double loop, and (3) deutero learning. In *single-loop learning*, the organization detects "errors" and makes corrections according to existing plans and policies. This is consistent with our discussions of contingency planning and classic PRM in earlier chapters (see, for example, Figure I.1 in the introduction to Part I). In contingency planning, the organization creates policies consisting of contingent actions, should "events" arise, and then implements these planned actions if and when events *do* arise. While specific actions might change, no modification is made to the underlying policies and plans that generate these actions.

In *double-loop learning*, when an organization detects errors, it takes action to correct these errors in ways that involve the modification of the existing plans and policies. The term "double-loop" implies a correction not only in response to errors but also in how the response is made. This is consistent with the type of learning we will be discussing in this chapter. As the organization modifies its project plan in response to acquiring new information (as it "detects errors"), it is creating new policies and implementing these new policies as it proceeds. Thus, actions, and the policies and plans that generate these actions, are changing over time.

The final level of learning, *deutero learning*, involves changing the learning system by which organizations detect errors and take action. In order for an organization to move from contingency planning to project learning, it will have to change not only what it does but also how it goes about doing it. Thus, the organization will have to shift from a project mind-set, infrastructure, and governance geared toward contingency planning to one geared toward active project learning. This level of learning will be addressed in Chapters 8 through 10.

5.2.2 Double-Loop Learning through Improvisation or Experimentation

So how do organizations carry out double-loop learning? In other words, how do they go about acquiring new information and how do they respond to this information, once acquired? There are two types of double-loop learning that organizations can undertake, which increase in the amount of effort required and responsiveness achieved: improvisational learning (learning in real time from action variations) and exploratory or experimental learning (stretching from trial-and-error to purposeful experimentation).

Improvisational Learning
In improvisational learning, real-time experience drives novel action at the same time that the action is being taken.[8] That is, planning and execution occur simultaneously, typically in response to problems or opportunities created by unanticipated events.

Improvisation can be in the form of new behaviors or product features, but they can also be in the form of new interpretive frameworks; problems can become opportunities when seen in a new light, thanks to changes in how the "problem" is framed.

The metaphor of improvisation comes from the arts. Take jazz, for example: Each musician varies his or her tune around a common and stable plan of rhythm and harmony. While the musicians vary their tunes spontaneously, in real time and in unplanned ways, they must strictly adhere to the rhythm of the ensemble. This limits the range of variations that are possible; thus, creativity is limited. In other cases, the time and coordination pressure is not so high, and so more creative variations can be sought. Painters, for instance, while loosely collaborating, work alone and can widely vary their individual styles.

Thus, improvisation can comprise a differing mixture of spontaneity and creativity. As the degree of uncertainty increases, in particular, as unk unks and complexity become important, creativity of improvisations becomes critical in order to be able to respond to the unk unks as they emerge.[9]

In jazz improvisation, spontaneity is high, as the jazz performer is creating in real time, but uncertainty is low, and unk unks nonexistent. The basic "plan" is known to all performers, and improvisation is mostly ornamental. This also applies to what we referred to as "residual uncertainty" in Chapter 1. In the PCNet project, the basic plan was known and the risk management office (RMO) was simply there to respond to unknown unknowns as they arose. The team had to be improvisational in that they had to respond to unexpected events quickly, but mostly this improvisation was driven by spontaneity to time pressure, rather than open creativity.

Rapid response to events is also critical in control-and-fast-response situations. With complex projects, it is critical to stay on a known control path. Any deviations must be dealt with quickly, often requiring individuals to simultaneously develop and implement a solution. However, again we see that unk unks are relatively minor as long as one stays within the known region of control. Improvisation is still based more on spontaneity than on creativity; the overriding goal being to keep the project within the known region of control. As a case in point, when the PCNet project management team discovered that local changes to the e-mail system created unanticipated system instability, they quickly moved to preempt this behavior at all local installations, thus keeping the system in its stable configuration.

When uncertainty is higher and becomes unforeseeable, improvisation must vary more widely, incorporating more on creativity than on spontaneity. Austin and Devin (2003), in a collaboration between a technology management professional and a theater professional, pointed out a very close parallel between managerial "knowledge work" and the work of artists, a parallel that is underutilized by project and new product development organizations. Describing a theater production and a software development project, they show how creative improvisation around a theme is necessary not only to react to external changes but also to *create* a superior solution that could not be foreseen or planned because the project is too complex—too many variables interact in determining how good the solution is. The authors see four qualities that make for successful improvisation of a team:[10]

1. *Release.* A method of control that accepts wide variation within known parameters. "Release" contrasts with "restraint." This quality is essential for the other qualities to work. It allows the creation of variety, which is the core of creativity.

2. *Collaboration.* The requirement that each party, in language and behavior, treats the contributions of the other parties as material to create with, not as positions to argue with, so that new and unpredictable ideas emerge.

3. *Ensemble.* The quality exhibited by the work of a group in which individual members relinquish sovereignty over their work and thus create something none could have made alone, a whole greater than the sum of the parts.

4. *Play.* The quality exhibited by a [theater] production while it is playing to an audience, or by the interaction among members of a business group, and ultimately between the group and the customer. This quality builds upon intimate interaction between the group and the audience, or customer, so the customer closely relates to the production.

These qualities capture the essence of improvisation; it means creative variation within a known range. When uncertainty and unk unks are high, a project team may "find [itself] frustrated by the simultaneous pressure to act and the inability to understand what is going on around them."[11] Here, a one-off improvisation is not enough to resolve the uncertainty and to find a good solution. Indeed, Austin and Devin emphasize that improvisation must be *embedded in a systematic process*, which essentially repeats cycles of creativity. This is what we describe next, as experimentation.

Experimentation

The most basic form of experimentation is trial-and-error learning, in which the organization develops and implements its plan but then closely monitors the situation to constantly evaluate whether and how the plan should be modified. The more systematic and exploratory experimentation becomes, the more it contains the *purposeful* search to uncover unk unks, without regard to the success of the trial.

The basic building block of experimentation is the Plan-Do-Check-Act (PDCA) cycle, often seen in operations (see Figure 5.4).[12] The key success factor for learning is to keep the PDCA cycle small and fast. That is, "failures" should come early and often, before they become catastrophic. Changes early in the project are always less costly than those that come late in the project. The ability to create early, small failures runs contrary to traditional project management mind-set, infrastructure, and governance. Many projects fail to implement the PDCA cycle because they use a model of project management where all tasks and requirements must be defined *before* the project can even get under way. The standard response is "How can you manage a project if you do not know exactly what the project involves?"

Significant up-front analysis and design are undertaken at great expense in terms of time and money, even in projects where there is no known feasible, let alone optimal, set of activities. This is necessary because without planning, the team has no basis on which to stand. However, by carrying out extensive risk analysis, creating risk lists and contingency plans, a team can easily fool itself into thinking that it has a robust plan. This is fiction under the presence of significant unk unks because the plan itself is not really known. The best that can be said is that the plan under consideration is simply a starting point for what is known about the project (we already touched upon this problem in Chapter 3, when we called PRM "double-blind"). Too often, the initial project plan, while flawed, becomes the new objective, while the original objective can get lost in the confusion. The project plan is seen as an end in itself, and no longer a means to achieve an end.

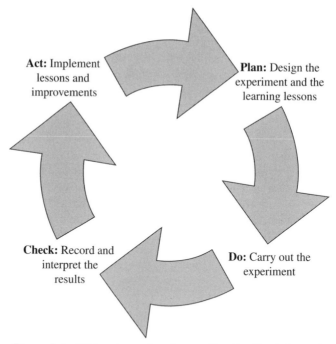

Figure 5.4 Trial-and-error learning as a Plan-Do-Check-Act cycle

Organizations must be prepared to make early mistakes, and make them often. The project should not be seen as a "single monolithic entity that is either frozen or liquid, but rather a more complex structure," one whose specifications can be progressively frozen as the project progresses.[13] Project commitments should be made piecewise as more information is revealed, rather than all in one go at the start of the project.

5.2.3 The Trade-off between Learning and Progress

In the extreme case of exploratory experimentation, activities are undertaken that look inefficient. The critical question is not "What are the next actions in an optimal project plan?" but instead "What are the next actions that will generate the most information about the unknowns in the project, and how can we best incorporate this information into our project?" Early actions are taken not so much as part of an initial project plan, but more a part of a learning process as in the design of experiments. At the opposite extreme of improvisation, exploration purposely maximizes variation in order to create the greatest opportunity for learning, not the greatest opportunity for immediate success. This is summarized in Figure 5.5.

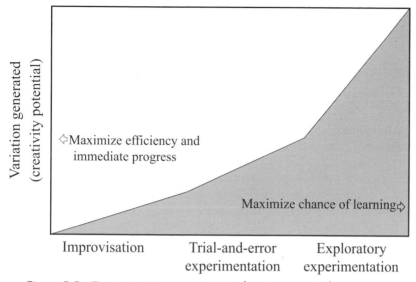

Figure 5.5 The trade-off between variation (learning potential) and progress

The more experimentation and learning the project team undertakes, the less efficient progress toward the targets becomes. However, when the plan is a fiction, progress is a mirage. The more important unk unks are, the more the trade-off is biased toward learning. And the earlier experimentation and learning can happen in the project, the lower is the efficiency loss, because early actions are usually cheap. Thus, experimentation should take place as early as possible.

In the presence of unk unks, up-front market analyses or technology performance analysis may be impossible, as critical factors affecting performance are unknown. Early experiments, generating early failures, rather than analysis, are critical for learning.

5.3 Back to Escend: Drawing the Lessons
5.3.1 Interpreting the Escend Example
The Escend Technologies story is an excellent example of what to do and what not to do in managing projects with significant unk unks. At the beginning of the story, we saw a management team attempting to utilize a planning approach to managing the project. As unk unks arose, which was inevitable given the nature of the project, the management team was faced with the pressure to act, but without fully understanding what was going on around them. They tried to improvise around a plan, but unfortunately, the plan was simply too unrealistic to offer much direction. Thus, the team became confused and began to focus everything on making the plan work, forgetting their original objective of making money from a market opportunity.

It took a crisis to bring the board around to recognizing that the original approach to the project was flawed. A change in mind-set was required. It became a priority to understand what they knew about this market opportunity and about Escend's capability to take advantage of this opportunity. The business plan came to be seen simply as a starting point, a way of asking critical questions. The important questions concerned what they knew and what they needed to know, and where the gaps were that needed to be explored.

The management team then began to systematically explore the unknowns. They could not design laboratory-type experiments—they were not in a powerful position to run trials with their clients. But they could, and did, systematically pose questions and make trial proposals to the market, as well as run technical tests with the product, in order to explore the nature of the market. In this way, Escend radically evolved its business model, not only by improvisation but by systematically exploring what it did not know.

The resulting business proposition, as presented in Figure 5.3, stands in stark contrast to the confused and ambiguous proposition presented in Figure 5.1. This is the result of a systematic approach to exploring what the team did not know, improvising within the current plan in response to unfolding events, and implementing a PDCA cycle where the results of each action are diagnosed for possible learning opportunities, which then change the business plan and future action.

Implementing a learning strategy in any project is not easy. It is human nature to want a plan and milestones by which it is to be measured. However, as we have seen in both the Escend and Circored examples, a project plan should never be seen as an end in itself: It is always the means to an end. The ultimate objective, whatever it is, should always be at the forefront of the project team's mind.

5.3.2 An Experimentation Process

The learning process diagram in Figure 5.6 summarizes the discussion so far.[14] The heart of the process is the experimental PDCA cycle, with embedded improvisation to maximize variety and, thus, creativity and learning potential. The "act" step emphasizes that experiments are not one-off activities but must be continuous in novel projects. The "play" position emphasizes that the evaluation of an experiment risks being insufficient if performed only by the team. Customer input is important to add fidelity to the experiment, but also to build customer intimacy, helping both customer and team to understand the other side better.

The individual PDCA cycle is embedded in a *process* of a stream of learning events. The principles of the process are, first, to accept *failure* as a source of learning. This is counter to the instincts of many managers who interpret every instance of something not working as a mistake. However, in novel projects with unk unks, you only know whether you have reached the limit of current performance when something fails. A failure is not a mistake; a mistake is a failure that produces no new information for the team.

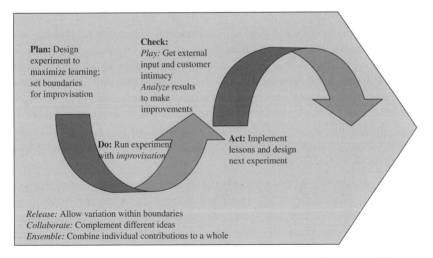

Plan: Design experiment to maximize learning; set boundaries for improvisation

Check:
Play: Get external input and customer intimacy
Analyze results to make improvements

Do: Run experiment with *improvisation*

Act: Implement lessons and design next experiment

Release: Allow variation within boundaries
Collaborate: Complement different ideas
Ensemble: Combine individual contributions to a whole

Design principles of the process:
- Recognize failure as a learning opportunity
- Experiment as early as possible
- Organize for frequent and rapid experimentation
- Integrate multiple experiment technologies

Figure 5.6 An experimental learning process

Second, information is most valuable if it is gained early. Chances of early success are small, but the costs are almost always lowest at the beginning, and opportunities for learning are great. Thus, the organization must *anticipate and exploit early information* if it is to benefit from early probing. If early experience cannot change later actions, the organization cannot benefit from experimentation. The important questions must be "What can be done differently?" The sooner one finds that out, the better for the project.

Third, the project team must be organized for rapid experimentation. For example, if the person, or group, who plans the experiment is in a different department than the one who executes it (and perhaps the person who evaluates the results is in yet another department), rapid turnaround becomes elusive because one party may not learn immediately when the previous step is complete, because priorities may not be aligned, or simply because the experiment enters a queue (an inbox on someone's desk!) every time it is handed over. Or, subteams may not have the incentives to share "failures" with other subteams. As a result, the team might not be able to execute quickly enough or to draw the right lessons from the results.

Finally, the organization should *combine multiple technologies* in order to maximize variation, or the opportunities for learning. The best "technology" of learning changes as the project progresses. At the beginning, learning may be possible from graphs, pictures, or customer questionnaires. Then, as more information is available, mockups or realistic renderings (simulations, pictures) may be better. At some point, partial prototypes and then system prototypes are needed in order to gain additional information.

The more exploratory the experiments are, in other words, the more specialized they are toward gaining information without contributing any "progress" to the current version of the plan, the more difficult they are to implement: (1) They are off-line—that is, early probes are designed more for learning than for success; (2) they require the team to "fail" early and often, something most project management organizations are not well suited for; and (3) they require rapid and effective learning from failures to effect change in the project. Exploratory experimentation is an iterative process of successive approximation, something quite alien to most project management organizations.

For exploratory learning to work, the organization must constantly ask itself several questions: "What do we know, what do we need to know, and what might we not know that we do not know?" "How do we best go about finding answers to these questions?" and "How do we best incorporate what we have learned into the project plan?" As we have seen in the Escend example, these are difficult questions for most organizations to ask.

In the face of significant unk unks, an organization can influence the technology used to make the project flexible, so that as events arise, changes can more easily be made at all stages of the project. Project flexibility provides the ability to modify the project in response to new information as the project progresses.[15] A project's flexibility can be increased in many ways, one of the most significant being project modularity. The more modular a project's architecture, the more one can change one aspect of the project without affecting other aspects. Technology choice can drive the underlying physical or informational architecture. A modular technology architecture allows changes in one module of the technology without significantly affecting other modules, in terms of either the physical (space, power, heat, etc.) or the informational interfaces.

Organizational processes can also have a significant impact on project flexibility. Classic stage-gate processes, while useful for classic projects where one can do up-front analysis to achieve at least a feasible, if not optimal, project plan, are poorly suited to projects with significant unk unks. Once the initial stages of the stage-gate process are completed, there is strong organizational pressure for everyone to support the project plan. When the project plan turns out to be infeasible, it is difficult to be the first to say so. In fact, it is human nature to disregard alternatives once a plan has been accepted and agreed upon. The more formal the stage-gate process, the more buy-in one gets into the resulting plan, and the more difficult it becomes to explore alternatives or to even recognize those events that are strongly indicating that the current plan may be suboptimal or infeasible.

Finally, stakeholder management is critical to maintain project flexibility. Uninformed stakeholders might put up with early changes to the project, but later changes are typically seen as problematic. Stakeholders must see the project the way the project organization does; they must be prepared for what is to come and they must have a way to measure progress.

Instead of seeing constant design changes to the agreed-upon project plan, they should see progress on learning about the ultimate solution as it emerges, or on progressively freezing the project specifications. This takes significant reframing of the project in the minds of the stakeholders. These issues will be addressed more fully in Part III.

Endnotes

1. For example, the learning organization has been defined as one "skilled at creating, acquiring, and transferring knowledge, and at modifying its behavior to reflect new knowledge and insights" (Garvin 1993, p. 80). More generally, learning has been defined as experience generating a systematic change in behavior or knowledge (for example, Argote 1999).

2. Section 5.1 is again based on Loch et al. 2005.

3. Cited from Crossan et al. 2005, p. 134. We will discuss a more systematic view of Escend's situation in Section 5.2.

4. Drucker 1985, p. 189.

5. The open-end process worked because Elaine was the person in charge. When she became CEO, Escend's morale was so low that no amount of cajoling was effective. After the layoffs, Elaine got the five remaining employees to make decisions by consensus (via daily and weekly meetings), which created a cohesive team that accepted both responsibility and accountability.

6. This hypothesis was confirmed in August 2004 by a report from industry experts that Escend commissioned for $40,000. Elaine concluded that "experts are good at messaging what you already know, but not at what you don't know." So she convened the board to brainstorm about unk unks for other vertical markets.

7. The typology was created by Argyris and Schon 1978, in their seminal book on organizational learning.

8. This definition is taken from Miner et al. 2001, p. 316.

9. See Crossan et al. 2005. These authors go as far as recommending that an organization should shun planning and only improvise. This is wrong; without a plan, the team has no foundation to stand on. Improvisation is useful only if it happens within well-defined boundaries.

10. See Austin and Devin 2003, p. 15f.

11. Crossan, et al. (2005): 134.

12. Thomke 2003 has proposed a modification of the standard PDCA cycle for prototyping in product development projects: (1) design the experiment, (2) build a prototype, (3) run the experiment, (4) analyze the results. This modification emphasizes, first, that learning typically happens on prototypes, or incomplete approximations of the final output, and second, that the analysis step at the end is critical to ensure learning and increased understanding.

13. See Thomke and Reinertsen 1998.

14. The diagram combines the modified PDCA cycle from Thomke 2003 with process principles from Thomke 2001, and the embedded improvisation as proposed in Austin and Devin 2003.

15. This concept is based on the concept of development flexibility offered by Thomke and Reinertsen, ibid.

6

Multiple Parallel Trials: Selectionism

In Chapter 3, we argued that there are two approaches to project management in the face of unknown unknowns that complement the traditional approach of PRM. In Chapter 5, we described the learning approach and explained how it can be applied. In this chapter, we explore the other fundamental response to unk unks, the one that we call selectionism: trying out several plans simultaneously and seeing, ex post, what works best.

Conceptually, selectionism is easy to understand. Here is a simple definition:

> In the face of uncertainty, one launches several solution attempts, or sub-projects, each with a different solution strategy to the problem in hand. If the solution strategies are sufficiently different, one would hope that one of them will succeed and lead to a successful outcome. Success depends on generating enough variations so that ex post, we obtain desirable results.

Let us use a thought experiment to explore what this definition of selectionism implies. Imagine that a military plane has crash-landed in the jungle of Laos, with a nuclear bomb on board. The bomb must be retrieved as quickly as possible, but the terrain is so complex that intelligence is not able to provide good tracking. Moreover, there are guerillas in the vicinity, so the search parties will have to shift dynamically, depending on where they encounter resistance. The military leadership decides to send seven search parties in parallel. Each can earn a huge reward (both monetary and in the form of a Medal of Honor) if they successfully retrieve the bomb. But only one will find the bomb and "win." The others will "lose." Their effort is still necessary, of course, because no one knows in advance which team will find the bomb. This problem has one, and only one, useful target of the search—namely, the place where the bomb is located. Any other place in the jungle is utterly useless for the purpose of this search. Thus, this kind of problem is called a "treasure hunt."

In projects, there is not usually a single useful solution; there are, rather, many system configurations that work more or less well. There are many solutions with higher or lower levels of performance. Finding the best is the challenge.

The challenges of implementing what seems a simple conceptual approach are quite daunting. Does one have the resources to launch several projects? Does one have the mechanism to identify early on the cul-de-sacs? What does one do with the teams that were committed to those cul-de-sacs? How do you remotivate them and keep them on board? And how does one enable the organization of learning the knowledge that a failing subproject often creates? Nature is a master at selectionism. It has a brutal way of getting rid of the failing projects: only the fittest survive. Nature also has a quite sophisticated learning system through genetic selection. Organizations cannot apply the same brutal approach, and need to find sophisticated answers to implement selectionism.

In this chapter, we first illustrate selectionism, and with the example of a small telecommunications equipment company that needed to completely redefine its business model, we will provide a more detailed illustration. Then we explain in more detail what selectionism is and where and how it has been applied. We conclude with a number of suggestions for how to implement selectionism where it is appropriate.

6.1 Selectionism at Option International
6.1.1 Company Background

This section gives an interesting example of how selectionism helped a small telecommunications equipment company to redefine its business model. Option International[1] is a small Belgian company created in July 1986 by Jan Callewaert, an almost prototypical example of an enthusiastic young entrepreneur.

The company was founded to design and manufacture small modems for portable computers. In the early 1990s, riding on the success of GSM (Global System for Mobile Communications) in Europe, it successfully developed a series of modem cards to enable PCs to be connected through a mobile phone with the Internet or private data networks. Small and entrepreneurial, the company decided to distribute the cards through its own networks, assemble the cards in house, and develop its own brand. The business model the company developed was that of a niche player in the end market (both consumer and business) for computer peripherals.

In 1993, Option International scored its first big success with the first PC card modem with pan-European-type approval.[2] In 1995, the company received a great deal of visibility with the launch of its GSM-Ready PC card that combined GSM and PSTN (Public Switched Telephone Network) modem functionalities in a single device. Its success was based on its technological strength. The company's technical team had a thorough understanding of radio technology and knew how to combine it with the requirements imposed by the PC manufacturers. Although in both radio technology and computer technology there were formidable competitors, Option International understood how to leverage its ability to combine these two technologies in a niche market.

Based on its technological prowess and its initial market success, riding on the Internet boom that made all small entrepreneurial high-tech start-ups attractive, counting on an exploding market (the integration of the PC market with the fast-growing mobile communication market), and intending to professionalize its management, Option International successfully floated itself on the Easdaq, the European equivalent of the Nasdaq, in 1997. There was cash; there was a proven business model; there was technological know-how, enthusiasm, and a booming market. How could this go wrong? But soon the clouds in the sky darkened.

6.1.2 The Challenge: An Unpredictably Changing Market

Being a niche player was an advantage, but also a curse. As a technological leader, Option International knew how to sell to early trendsetters, but as soon as a larger market for modem cards opened up, the larger electronic component manufacturers became interested, and the computer manufacturers wanted to embed the functionality within the PC. It became clear that Option was a company with one product, and a product with a short life cycle, moreover. Being successful required constant renewal in that product and staying ahead of the potentially big competitors with strong manufacturing capabilities and large distribution networks.

Moreover, the market became complicated. Hitherto, it had been easy: The consumer had a PC and a mobile phone, and Option was able to connect the two, basically whatever the brand of either. But the market started experimenting with alternative terminals. Mobile phone manufacturers, such as Nokia, launched smart phones that had PC capabilities. If this move became successful, the road warriors would not need to bring their PCs along and would work with a smart phone. No more need for a data card from Option.

In addition, personal digital assistants (PDAs) made inroads into some PC uses and functionalities. PDAs also needed to be connected to the GSM world, but many of the suppliers had proprietary system software. A generic product produced by a supplier like Option would no longer do the job.

Not all the market news was bad. At the end of the 1990s, people in many industries were weighing the advantage of connections over the mobile phone network against remote operators. What about isolated vending machines in places where there were no fixed line connections, that could provide a regular update over the mobile phone network on the evolution of the stock in the machine? Was that a market for a modem card provided by Option? Or what about a fleet of delivery vans where drivers would have a custom-designed terminal to fill in orders or deliveries, that would, over the mobile phone network, communicate with the computers at headquarters? Perhaps an interesting market for Option?

Technologically speaking, Option could design solutions for these markets. But they demanded fundamentally different business models. Rather than selling to the trendsetting (and relatively price-insensitive) road warrior who wanted to be connected to the Internet or his or her company's intranet all the time, Option would have to get into business-to-business marketing and sales. Development timing would have to be coordinated with the customer; sales volumes could be much higher, but so could pressure on margins. This would require a much better mastery of operational processes. And Option would probably lose its visibility toward the end customer. A marketing strategy à la Intel (Intel Inside) did not look credible for a small player like Option, and the big suppliers of PDAs would probably want to sell the modems under their own brand.

The uncertainty about the market was compounded by technological uncertainty. The world was full of rumors about the replacement of GSM by GPRS (General Packet Radio Service) and UMTS (Universal Mobile Telecommunications System). Operators were outbidding one another like mad for UMTS licenses. While the technological trajectory seemed relatively straightforward, at least in Europe, there was the wildest speculation about the timing of the introduction of the successive generations of mobile technology.

6.1.3 Selectionist Trials to Find a New Business Model

On the technology side, Option International pursued a learning approach, as described in the previous chapter. But the issue was still what kind of business model would survive. The direct relationship with the customer? The subcontracting role for major device manufacturers who wanted to embed communication capabilities in their end products? At this point, the company began to conduct selectionist trials.

First, there came an attempt to brand Option products as PC accessories. A logical evolution of Option's own products seemed to be some forward integration, for instance, producing a card to eliminate the mobile phone by turning the PC into a phone. Thus, a card was developed that had an antenna, could incorporate the SIM (Subscriber Identity Module) card of the mobile operator, and could enable (voice and data) communication via a PC. The product was launched in 1998 as the FirstFone. It was a first for the world. The same technology could also be used for a GSM adapter for a PDA. In the last quarter of 1997, the Snap-on, an adapter for 3Com's PalmPilot, had been launched. At the same time, Option International enhanced the GSM-ready card with additional capabilities, such as ISDN, and began to work on the inclusion of GPRS functionalities.

The initial market commentaries about the FirstFone were positive, but the product did not sell well. Did that mean that the direct sales strategy was wrong? Perhaps alliances were needed to sell the product. With this in mind, Option signed contracts with companies such as Omnitel in Italy, T-Mobil in Germany, and Telenor in Norway over the next few years, in order to increase FirstFone sales. While none of these contracts became blockbusters, some interesting lessons were learned. Option discovered that perhaps its product did not belong in a computer shop, or that it might be bought by an industrial client in an independent way. Perhaps the customer, private or corporate, did not see this as a computer peripheral but rather as a communication device, to be bought, like a mobile phone, through a telecom operator.

The Snap-on generated some sales but was not the expected blockbuster, either. Nevertheless, the market for embedded devices or proprietary add-ons to terminals remained intriguing. Option International signed a development contract for a wireless communication module with Handspring, the new PDA producer. Soon afterward, a contract was closed with Compaq to design and produce a GPRS solution for the Compaq iPAQ Pocket PC. Both products were launched within a year, in 2000–2001. Both times, Option had high expectations. It hoped that add-ons to PDAs would lead to a volume business and that this would help the company to make the transition from a startup to a mid-sized company. To be able to respond to the development and quality requirements of its customers, it had to completely revamp its development organization and its manufacturing approach. Development capacity was enhanced by the acquisition of facilities in Germany and the UK (Cambridge), and the production facility in Ireland was extended.

The second selectionist trial was an attempt at a supplier role to telecommunications companies. From making communication add-ons to PDAs to viewing oneself as a development contractor in the communications world was only a small step. Why not build phones around the radio device that Option had already perfected? The question was triggered by a Chinese electronics manufacturer who wanted a fully designed mobile phone from Option. Accepting this contract did not make sense when Option defined itself as a producer of communication cards distributed under its own brand. But as an R&D subcontractor, which it had already become for Handspring and Compaq, it *did* make sense. So the contract was accepted.

None of these subcontracting relationships proved to be very successful. For the Chinese contract, a significant amount of the investment had to be written off in 2002. The OEM contracts with Handspring and Compaq[3] *did* deliver significant revenues, but the relationships remained limited to one generation of the product. In 2001, 94 percent of revenues came from these contracts. Option learned many lessons from this experience, two of which were significant. The market was clearly moving toward embedded products, and the survival of an independent Option brand in the market was unlikely. Second, sales had remained significantly below original expectations, and independent observers of the communication card market had concluded that computer and PC manufacturers had insufficient knowledge to sell communication solutions.

Thus, the first two experiments in defining the business model were unraveling: independent sales of branded Option modules as computer periphery, and design and production for telecommunications OEMs. Yet, a third experiment was ongoing. Option had incorporated GPRS into its own "in-house" card and launched it as the "Globetrotter" in the first quarter of 2001. Now it was looking for the successor, which was to include a 3G solution. Option International could have pursued two market approaches in parallel: the development of an own card as well as the development of add-ons for terminal producers. But a new option came on the radar screen.

Some telecom operators did not have a good understanding of what consumers would do with GPRS and UMTS. It was clear that voice communication did not need these technologies and that they could only make money if consumers used the mobile networks heavily for data communication. Thus, data communication had to be stimulated. The network operators gradually realized that the use of communication cards also needed to be stimulated and that these cards were really essential to the rapid rollout of GPRS and UMTS. As a result, in the first quarter of 2002, Option was approached by Lucent, a telecom equipment manufacturer, to codevelop a high-speed data solution for UMTS networks. In the third quarter of 2003, a direct collaboration with Vodafone was announced. Finally, this trial seemed to work. In both 2003 and 2004, Option International doubled its revenues and returned to a healthy profit, after several years of losses.

This example illustrates how this small company followed a fairly straightforward trajectory with respect to its technology, but had to develop a totally new business model. The business model had to be redefined because Option International was confronted with enormous market uncertainty in 1997 to 1998, combined with unk unks created by the Internet turbulence around 2000, as well as by the erratic behavior of the telecom operators when it came to UMTS licenses. If it had had time and cash, Option could have followed a learning strategy. But it had neither. It had no choice other than to continue with its original business model of selling computer peripherals under its own brand, to generate immediate sales. In addition, it experimented with OEM relationships and pushed this even to being a telecommunications R&D subcontractor.

Not all these experiments were launched at the same time, but they were seen as true alternatives, not as "competing" but as feasible options that needed to be explored in order to make an informed choice. At all times, the knowledge gained from one experiment was immediately applied to the other experiments. What was learned from the FirstFone in the market was applied to the partnerships with the operators. What was developed in terms of insights by working with terminal manufacturers was shared back into the projects with telecom operators.

Collaboration with telecom operators has now emerged as Option's superior business model. Each of the multiple selectionist experiments was partially successful, and through good communication among the various teams, a robust business model was developed. As we write this book (in the fall of 2005), the company seems to be fully committed to the further development of its partnerships with a wide range of operators. At the end of 2004 and the beginning of 2005, there was almost one announcement per week of a partnership with an operator.

The process of identifying the best business model was anything but smooth. A cursory analysis of the annual reports and press releases of the last five years immediately dispels any illusion of straight "progress." The company went through rough times financially, and press releases over the period 1999 to 2002 show repeated reorganizations of sales and marketing (directors and salespeople were hired and fired), of the development capacity (a laboratory was acquired in the UK and Germany, but the one in the UK was fairly rapidly reduced and then closed down), and of operations (production was outsourced to Jabil, a well-known contract manufacturer).

6.2 Explaining the Principles of Selectionism

The idea of trying multiple solutions in the same way as sending multiple search parties into the unknown jungle in order to choose the best attainable solution among them afterward is captured in the image of a rugged landscape. This refers to a solution "space" of design parameters with performance peaks and valleys. A peak corresponds to a project parameter constellation that yields high performance. The problem is to find a "good" peak of sufficient performance. In complex problems, there are many good solutions (and even more bad ones), and optimization is not possible; one cannot hope to find the "best" performance through analysis and incremental search. "Good enough" is what counts. Figure 6.1 shows parallel search parties sent into the mountainous terrain of a parameter space to find a good peak.[4] We will further develop the performance landscape metaphor in Chapter 7.

Thus, selectionism combines an overarching goal with the individual ambition of each search party. This clarifies the challenge: It is about balancing individual rewards (to get the teams to try hard and take personal risks in order to succeed) with a group objective to find the best possible solution for the organization. What matters is that the bomb in the jungle is eventually retrieved, not by whom. If the groups overemphasize the personal targets, they will not collaborate, or support one another, or share information (they may even hide information), and they will thus put the overall mission in jeopardy. But if they do not feel that they will be rewarded for success, they may not give their best.

This has important implications for the management systems used to lead the project (what we call "project infrastructure" in Chapter 9). The success of the individual project is subordinate to the success of the overall mission. Indeed, management wants to stop an individual trial when it becomes clear that a different project is doing better. From our example, as soon as there are indications that one or several of the teams have got close to the location where the plane crash-landed, you do not want to expose the other teams to the dangers of an unknown terrain.

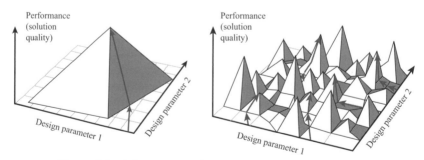

Figure 6.1 Selectionist searches in configuration parameter space

However, the people on the stopped project should not be punished if their project stops and a different one succeeds—it is the nature of the ambiguous and complex overall task that it was not foreseeable who would find the prize. All teams are necessary and make a contribution. Information must be shared, and all teams must share in the success of the overall project. We will further discuss these questions of project infrastructure in Chapter 9.

The treasure hunt example is obviously simplistic and artificial. But as we have seen at Option International, companies follow similar approaches when confronted with complexity, or unk unks. They often do so reluctantly; it is not easy to juggle several different (sub) projects with the same or a similar goal.

One study of 56 new business development projects suggests that one key difference between firms that are able to adapt to a changing environment and those that fail to do so lies in their ability to engage in selectionism,[5] creating a variety of different solution approaches. Pursuing improvement of the solution that one has may lead into a competence trap by settling on a solution that is subsequently shown to be inferior. As the rate of environmental change accelerates, in other words, as the degree of unk unks increases, selectionism increases in importance and produces better solutions than "continuous improvement." This is consistent with findings in systems engineering methods: The more complex a system is, the less the engineering team is able to find a "best solution," and the more important it becomes to try multiple different starting solutions to have one of them lead to a "good" solution.[6]

Dorothy Leonard-Barton (1995) describes a version of selectionism that she calls "Darwinian selection." This refers to experimenting with multiple models in the market, simultaneously. For instance, Sharp Corporation introduced several PDAs in 1993 to 1994. One was based on a proprietary operating system, sold in Japan as part of the Zaurus product line. Second, Sharp licensed Apple's Newton architecture to produce Expert Pad, as a direct competitor to Apple. Third, Sharp introduced a PDA based on GEOS from GeoWorks. In addition, it had plans to launch products based on Microsoft's WinPad and any other operating system with a chance of ultimately dominating the market. Products that don't do well in the market are mercilessly culled.

Such Darwinian selection seems to be more popular with Japanese companies, compared to the United States, where companies are less comfortable with this throw-it-against-the-wall-and-see-what-sticks attitude. Several hundred brands of soft drinks are brought out annually in Japan, most of which fail within a year. As one puzzled Western executive said, "These guys must be throwing money away; I've never seen such an unsuccessful bunch of marketers." However, until the early 1990s, Japanese consumer goods manufacturers were able to use this approach to severely curtail the success of Western companies. Although many trials failed, they offered learning lessons, and the successes opened new market niches so quickly that the targeted marketers had a hard time to react quickly enough.[7] Still, Darwinian selectionism is very costly, especially when the resource mobilization to pursue each one of the independent trials is high. Indeed, many Japanese companies seem to have pulled back from Darwinian selectionism in the mid-1990s.[8]

Generating variety needs to be balanced by ex post selection in order to be productive. In the case of Sharp's PDAs, the selection was indeed Darwinian, which means that the end customer market selected the fittest product. This is an expensive approach, as the market launch is the most expensive phase of any product development project. Therefore, many firms try to generate variations at the prototype level and narrow them down to a final solution before market launch.

For example, Toyota builds many prototypes of a new car, broadly considering sets of possible solutions to a design challenge and gradually narrowing the set of possibilities to converge on a final solution.[9] From the start, there is a wide set of possibilities, a wide net as they call it, followed by a gradual elimination of weaker solutions, based on additional information emerging from development, testing, customer input, and the achievements of other sets. As car designs converge, the different participants commit to staying within the set(s). According to Toyota, this form of selectionism, combined with a gradual refining of the space of sets, makes finding the best solution more likely. It takes more time, early on, to define the space of solution sets, but once the different trials have been determined, Toyota seems to be able to move more quickly to convergence and to production, than by using more traditional forms of concurrent engineering.

Implementing this set-based approach is not a simple management task. It requires the presence of a strong chief engineer who acts as system architect and oversees the process of narrowing down. This ensures three conditions of success:[10]

1. The development of the acceptable set space—that is, the definition of the constraints and criteria that characterize acceptable sets. An important part of this exercise is the constant communication about sets of ideas and regions of the design space, not about one idea at a time.

2. Integration of achievements from the sets that are developed in parallel by looking for intersections of feasible sets. In this integration exercise, it is extremely important to look for robust solutions—that is, solutions that can be successfully applied under many different combinations of constraints.

3. Confirmation that solutions are feasible before committing to them.

It is striking that in the description of the methods used by Toyota, it is explicitly mentioned that while in normal project management " . . . design processes [are viewed] as networks of tasks and [are controlled] by timing information handoffs between the tasks, as in the familiar PERT chart, Toyota views its process as a continuous flow, with information exchanged as needed." This stresses the need for a fundamental characteristic for selectionism. Selectionism is not a war among competing project teams that do not communicate or share knowledge with one another. Rather, it is the organization of multiple teams that constantly communicate with one another, share learning, and enrich one another.

6.3 What Makes Selectionism Work?

The comparison of the case examples above enables us to highlight some of the principles of selectionism. One can classify them in the following five (not fully independent) categories formulated as questions:

- ▲ In what space are we going to form alternatives? What is the set space of feasible and practical solutions?
- ▲ How many options, sets, or experiments can one afford to carry out simultaneously?
- ▲ When do we stop options or projects?
- ▲ How does one ensure that the selection indeed happens, and how does one create the commitment to the selected outcome?
- ▲ Key to the success of selectionism is the ability to integrate learning across the projects. How does one leverage the learning or other benefits from the nonselected experiments?

6.3.1 In What Space Are Alternatives Developed?

Selectionism is about creating variety and learning from it. However, the variety must not be unlimited. The options we want to pursue must fall within a feasible and practical space.

For example, on the subject of its set-based concurrent engineering, Toyota clearly states that success depends on carefully determining the set space. At the outset of the project, they strongly emphasize a correct definition of the design space, within which the sets have to fall. "Map the design space" is how Toyota characterizes the set of alternatives used in its convergence process. Functional departments within its development system simultaneously define feasible regions from their perspective. Each department, such as body engineering, chassis engineering, or production engineering, determines, in parallel, the primary design constraints on the system based on past experience, analysis, experimentation, and testing, as well as on outside information. These design constraints are translated into engineering checklists, which are used throughout the project to filter possible trials or sets.[11]

Option International, too, was tightly restricted in the selectionist trials that it could consider. Remember that its search for a new business model coincided with the heyday of the Internet boom. The temptation to opt for independent distribution of PC cards over the Internet may have been great, but it did not fit the observation that data communication cards also required close collaboration with mobile phone operators or terminal producers. Its variety space was also limited by its technology and the industry environment in which it operated.

6.3.2 How Many Parallel Trials?

The answer to this question is influenced by at least four drivers: (1) What is the cost of one trial or experiment; (2) how soon can one decide to stop a trial, (3) what is the degree of uncertainty, or unk unks, and complexity one needs to overcome, and (4) how strong is the managerial capacity of the organization to live with the ambiguity of the existence of parallel projects.

The more complex the problem (or subproject), the more trials are required in order to ensure finding a good solution. We will discuss this effect in more detail in the next chapter. But as was suggested by Leonard-Barton, in her description of Darwinian selection, the amount of resources needed to launch an experiment compared to the overall amount of resources the firm has available will obviously determine the number of experiments one can carry out. In the case of Option International, it was clear that the company could only pursue a limited number of development projects, given the size of its development department. Since each partnership or market approach required independent development, the company could only handle a limited number of contracts and market experiments. Even a giant like Toyota, while perhaps experimenting more than its competitors, is constrained by the number of sets it wants to, or can, coevolve.

Costs are not limited to resource commitments only. When the selection is visible to the outside world, such as in the case of Sharp or Sony, who both launched several versions of a product in order to figure out which one would get the best market response, there is potentially an important implication for the image of the company. If there are too many failures in the market, this may have a negative impact on the image of the organization. This effect is, to some extent, dependent on the tradition in the market. It is perhaps for this reason that selectionism is more often used by Japanese companies than by U.S. companies (see the discussion in Section 6.1). Traditionally, Japanese consumers have accepted that companies launch products in order to test their attractiveness. This acceptance level may be lower in the United States or Europe.

In the same way, a stock-market-listed company cannot be too daring in the array of business models it tries out. It needs to be firm in its communication with the financial markets.[12] Otherwise, its image (and stock price) will suffer.

Costs can be limited by restricting the duration or depth of the experiments. The earlier one can stop a trial that leads to a cul-de-sac, the better. The key to success is to make the impact of failing trials as small as possible. One can limit costs by stopping trials once one has a good idea that they are bounded by other alternative trials under way. We will come back to this in the next section.

The number of experiments is also dependent on the nature of the unk unk. There is a large literature in the field of combinatorial optimization that shows that the more complex a problem, the more parallel trials should be run. Thus, the more complex the problem, the more worthwhile it is to pay the cost of selectionism.[13] Higher uncertainty may also make selectionism attractive: If the team does not know the project terrain, it may choose to try several different solutions in parallel, hoping that one of them is appropriate for the environment that has emerged when the project is finished. Venture capitalists often view their investments this way.[14]

Finally, the number of experiments, perhaps even the capacity to apply selectionism at all, is influenced by the capacity of the project organization to live with multiple subprojects, of which several will be disbanded. As in the treasure hunt example from the beginning of this chapter, the organization wants to motivate the individual teams sufficiently so that they perform at their best, but it does not want to demotivate them when they are not "the lucky one" that finds the treasure.

The organization also needs a manager who can oversee a multitude of parallel projects. Not every manager has the skill of juggling many balls at the same time. In some cases, that person is called the program manager. In the case of a small company like Option, the role needs to be fulfilled by a person who is high up in the hierarchy. In this particular case, it was the founder and CEO himself who took it on his shoulders to carry through the project of the business model redefinition. In the case of Toyota, it was the chief engineer, the most senior technical decision maker on the team, who took on the role of system architect and lead designer of the vehicle. We will discuss the characteristics of this manager in more detail when we treat the project mind-set and the infrastructure in Chapters 8 and 9.

6.3.3 When to Stop Trials?

As we already mentioned in the previous section, the earlier one can stop a trial that leads to a cul-de-sac, the better. Costs strongly escalate over the course of a project, in many cases by a factor of 10 at each transition, from specifications to design, to development, to manufacturing launch, and finally to market launch.[15] While the cost escalation factor varies, obviously, from project to project, it is a good rule of thumb to bear in mind that the earlier one can stop a trial, the earlier one can select the appropriate course of action out of a number of experiments. As a result, the cost will be lower and the number of experiments that can be launched will be greater. The better one is able to organize the selection, the more variance one can create up front.

Selecting early on is highly dependent on the amount of communication among the different trial teams. If we go back to our artificial example of the teams searching for a nuclear device, it is clear that the faster one can recoup the information gathered by the different parties, the greater the chances are that one can narrow down the search area, and even take some teams off the operation.

In the case of Option International, there was always excellent communication among the different teams, in the sense that the CEO was heavily involved in each of them and could use the information collected in one experiment to improve the actions in another. As in the case of the chief engineer at Toyota, this shows that a strong overall leader, who has a good overview of all trials and is in charge of the process of convergence, is a second essential contribution to achieve selection early.

6.3.4 Ensuring Selection and Commitment to the Chosen Trial

Selection of the trials is essential. One of the biggest risks of selectionism is not to reach closure. Procrastination is the Achilles' heel of selectionism. While we stated the case for an early weeding out of dominated trials, we do not call for early selection at all costs. Converging too early can be risky. There is an optimum between letting the parallel trials run as long as necessary to get the benefits and cutting out some of the trials in order to limit costs. At Toyota, management insists on broad exploration to avoid convergence that is too fast and constraining. At Option International, the CEO, who had a complete overview of the different projects, paced both the introduction of new experiments and the conclusion of existing ones.

For a good process to come to selection, at least two components have to be in place: (1) excellent communication regarding the achievements of each parallel trial and (2) consolidation of the robust results—that is, the results that hold across different trials.

First, trial teams need to communicate with regard to positive achievements as well as failures. Moreover, evaluation of these results needs to be done in a shared way. The leaders of the trial teams need to sit down together to evaluate where each trial stands. Once again, selectionism is not served by ferocious competition between the trial teams. Such competition would lead to information hiding, which would be detrimental to all parties involved.

Second, robust results are those that emerge from different trials and hold under a variety of conditions. Let us return to Option International. In every business model that was explored, it became clear that the biggest beneficiary of a widespread adoption of PC communication cards was the telecom operators, because it was the only way they could stimulate data communication over the GPRS and UMTS networks. This was a robust result that still left Option a choice as to how to leverage this interest. Another organization with a stronger market position may still have attempted partnerships in which it could preserve its brand identification.

The robust result, however, implies that partnerships with operators should be a component of any solution.

Having selected an outcome has little value if, afterward, some team members switch back to alternatives or previous trials. Regression can be as dangerous as procrastination for the effectiveness of selectionism. Once an outcome has been narrowed down or selected, the team members must stay within the narrowing funnel so that other team members can proceed with their work, knowing that convergence has taken place.

Commitment to the solution depends on the degree to which members of the not-selected teams feel ownership of the final outcome. This can be best managed by attempting to enrich the selected outcome with contributions from other groups. This often improves the solution (as more information and expertise has been poured in), and if team members recognize their contribution in the selected outcome, they may adhere to it, even though it is not their own.

6.3.5 Leveraging the Benefits of the Nonselected Outcomes

The fifth principle for selectionist trials is that one needs to avoid the loss of the benefits of the nonselected outcomes. Benefits always include increased know-how and learning. In addition, there may be spin-offs, or an infrastructure that was created to explore the option and that can be used for other projects. We have three suggestions on how to manage this process of leveraging: ensuring common ownership of the end goal across the trial teams, fostering knowledge diffusion throughout the organization, and providing the teams with the appropriate amount of autonomy.

Often, the benefits of selectionist projects are embedded in people. The tacit knowledge they built during the experiment is valuable. Leveraging this value in the organization requires these people to remain motivated and to be distributed over the organization after the project's end, such that their knowledge will have the highest impact. As we have discussed, this requires giving all team members the perception that they contributed to the chosen outcome, and that they have a stake in that outcome.

Distributing the knowledge requires careful career management, which ensures that people move into new positions based on their performance, their potential, and the degree to which their tacit knowledge can help the organization that they join. The first two elements (performance and potential) are obvious. The third one is rarely applied but is essential if one wants selectionism to be effective for the organization.

Let us now turn to the issue of autonomy.[16] It has two components: goal autonomy and supervision autonomy. Goal autonomy has to do with the way performance goals are set. At one extreme (high goal autonomy), managers may give a group complete latitude in terms of goals, focusing on possibilities and opportunities. At the other extreme, managers may be directive, defining very specific goals and outcome criteria. Supervision autonomy refers to the specification and supervision of operational activities. A project group will have higher supervision autonomy if it has greater local discretion, permitting greater heterogeneity in day-to-day activities.

In order to increase the learning effectiveness of selectionist trials, one needs to increase both goal and supervision autonomy. Greater supervision autonomy allows for innovation in problem solving and helps reduce the strain on the team's information-processing capacity. It also provides an inducement for individuals to exercise greater individual discretion, leading to greater motivation and commitment.

The need to grant more goal autonomy is perhaps less intuitive. Traditional project management argues that clarity and measurability in the project goals are key factors for success.[17] The value of clear authority structures and working relationships for a project is seldom questioned. Project management preaches that goal autonomy should be relatively low in order to be conducive to good performance. As we discussed in Chapters 3 and 5, unk unks make plans, and their targets, just stakes in the ground. Thus, goal flexibility is valuable, both for the selectionist approach and for the learning approach.

6.4 Conclusion

In Chapter 3, we introduced two alternatives (learning and selectionism) to traditional PRM. In this chapter, we described and illustrated the selectionist approach. It consists of generating enough variations at the outset of the program in order to yield some desirable results ex post. It is not about creating a set of parallel projects that compete with one another to prove one another wrong. Selectionism is about creating variety and learning from it.

While it is a desirable goal, avoiding the negative effects of potential competition is not easy. Selectionism is quite difficult to manage. Summarizing our discussion, three principles stand out. First, the role of the program manager, the chief engineer, or the senior VP who oversees the set of projects is very important. Second, preserving open communication among the parallel trial teams is a condition for success. Motivating the members of the nonselected teams by giving them a stake in the final outcome is a third condition. We will look at all these elements from a different perspective in Chapters 8, 9, and 10 when we discuss the implementation of projects plagued with unk unks. But before discussing implementation, we need to see how to choose between the learning approach on the one hand and selectionism on the other. That is the topic of the next chapter.

Endnotes

1. More information can be found on the company's Web site, www.option.com. More details about some of the items mentioned in this case study can be found in the company's press releases or its annual reports.

2. Type approval is essential in the telecommunications world, where compatibility with the operators' networks is a necessary condition in order to be able to sell products. Without such approval, it is almost impossible to get access to the telecommunications networks.

3. There were other OEM contracts of a similar nature, but none of the same size as those with Handspring and Compaq. We keep them out of the story for simplicity.

4. The metaphor of a rugged landscape was popularized by Stuart Kauffman (1993) in the context of biological organisms performing well in the space of the genetic code. Rugged landscapes have also been used in project management literature (see, for example, Sommer and Loch 2004) and strategy literature (see, for example, Beinhocker 1999).

5. See McGrath 2001. She refers to selectionism as "exploration," or the generation of a variety of approaches.

6. As examples of systems engineering and numerical optimization, see, for example, Fox 1993 or Sobiezczansky-Sobiesky et al. 1998.

7. See the discussion in Jones and Ohbora 1990. They called Darwinian selection "product churning." The quote of the Western executive is on p. 21.

8. See, for example, Stalk and Webber 1993.

9. See Sobek et al. 1999.

10. See Sobek et al. 1999.

11. See Sobek et al. 1999.

12. The reports by financial analysts on Option, as well as its stock price are an illustration of this. During the period of experimentation, analysts were requiring more unambiguous information, and the stock price hit rock bottom. Since the period of experimentation, the realized revenues have been in line with announcements, and the stock price has increased by an order of magnitude.

13. For an example from numerical optimization literature, see Fox 1993.

14. See Sahlman 2000.

15. This has been shown empirically many times in new product development projects, for example, Soderberg 1989, or Terwiesch, Loch, and De Meyer 2002.

16. See McGrath 2001.

17. See, for example, Boddy 2002, Chapter 5.

Selectionism and Learning in Projects

We have introduced learning and selectionism as fundamental responses to the presence of unk unks, and we have demonstrated how they work in projects. But up to now, we have presented them in isolation of each other—we have not discussed under what circumstances each approach may be more promising, nor have we said anything about how they complement each other in a useful combination.[1] These are, therefore, the two goals of this chapter: comparing selectionism and learning, based on a priori identifiable project characteristics, and demonstrating how they can be combined.

We remind the reader that we want to propose a priori choice criteria for a combination of selectionism and learning, at the outset of the project, before unk unks have been revealed. At the outset, we can only identify knowledge gaps, or the potential for unk unks, but not the unk unks themselves. Therefore, the framework must be qualitative, a rough decision guideline.

Selectionism and learning can be applied at the level of an entire project, or at the level of subprojects, or at individual problems to be solved (we saw this in the diagnosis of the startup company, Escend, in Chapter 4). Clearly, selectionist trials of entire projects, even entire companies, are constantly conducted by markets. For example, one engineering provider may compete with proprietary technology, another with low-cost execution by "offshoring" the labor contracts, and a third may have access to attractive financing structures. The market chooses which approach is most successful. Competitors learn from one another by copying what works and avoiding the mistakes that they see others make. This is the essence of competition.[2] The same holds not only for projects but also for companies and their strategies in general.

Venture capitalist companies also apply selectionism at the level of entire startup projects. They take an explicit portfolio approach, looking for tenfold returns and killing the startups that do not perform. For the entire portfolio, returns of 20 to 40 percent are often achieved. It is not unusual for a large VC to have several portfolio companies that pursue the same target market. Learning occurs when each startup evolves as it goes through successive financing rounds, sometimes with substantial changes in its business model.

Large technology companies (such as IBM, Xerox, AT&T, and Siemens) have repeatedly tried to replicate the VC approach of project selectionism inside their organizations. However, for economic and organizational reasons, there are difficulties in applying VC-style selectionism to entire projects within established organizations. By definition, selectionism means that most of the trials are abandoned. In the eyes of controlling management, abandoning projects is inevitably synonymous with failure and wasted resources, and thus becomes increasingly difficult to justify when times are tough. Compensation in large companies is geared toward continuity of careers and fairness, while it varies much more in startups. VCs are focused on financial returns and ruthlessly kill projects that do not appear to deliver. In large companies, the continuity and strategic rationale of the business are almost impossible to ignore. As a result, project cancellation is harder in large organizations, which compromises the VC business model and, ultimately, may compromise both financial returns and strategic rationale.[3]

This is felt not only by large companies but also by the VCs themselves— their investments have turned increasingly to proven and more incremental startups since the burst of the bubble in 2001. Independent of the economic cycle, abandoning projects as trials is affordable only when the cost of each trial is small relative to the overall portfolio.

Thus, many companies that execute major projects find that they cannot apply selectionism at the level of entire projects alone but must also consider selectionism at the subproject level, and learn within each subproject how to improve the chances of success. In complex and highly novel projects, we typically see a combination of both approaches: learning and selectionism. Section 7.1 describes such a typical example. In Section 7.2, we discuss principles of how selectionism and learning complement each other, and how they might be combined. In Section 7.3, we reexamine the Circored example to illustrate how our framework could have been applied to this project. We present our conclusions in Section 7.4.

7.1 Selectionism and Learning at Molecular Diagnostics[4]

The example is a European pharmaceutical startup company that raised €7 million in 2002 to develop a new technology for a highly sensitive diagnostic of various diseases (the first round was intended to get the company to the proof of concept milestone). The technology consisted of identifying a modification of one of the four bases in DNA, cytosine, which may make the gene in which this base is included dysfunctional (this kind of modification of a base in a gene is called "DNA methylation"). Identifying a possible dysfunction in a gene via this marker is highly sensitive and can serve as a powerful early indicator.[5]

Fortunately, management was well aware of the unforeseeable nature of events that might hit them. They put several things in place to deal with unk unks. First, they instituted a "strategic pot of money," a (small) budget to deal with unplanned events, "just enough so we can quickly hire an expert or buy a license."[6] This played a role analogous to Metal Resources Co.'s residual uncertainty management described in Chapter 1.

The company also made explicit efforts to scan the market environment and diagnose trends as well as uncertainty. They conducted weekly patent screening and employed an academic advisory committee to learn about new technologies and bounce ideas; the input from the advisory committee also helped them to chart the initial course. They also built up intensive contacts with pharmaceutical companies and clinics to learn about potentially interesting products and technologies. This diagnosis effort did not come without cost—the patent screening alone cost them €400,000 per year. The advisory committee and industry contacts were cheaper but cost a lot of scarce management attention. It turned out that no patents forced them to fundamentally change their plans; they bought a few patents but essentially went on as planned.

The company used both trial-and-error learning and selectionist trials. They foresaw trial and error from the beginning, albeit only within a well-defined area: They were willing to experiment with the clinical indications to be diagnosed; essentially, all autoimmune diseases (such as cancer, diabetes, etc.) were considered fair game. But the diagnosis technology was fixed, namely, DNA methylation.

They conducted parallel trials by exploring several potential technologies to read DNA methylation patterns and narrowed them down to two.

Some did not work, and among the remaining candidates, they used comparative tests. Again, these parallel trials were anything but cheap, costing about €2 million per year.

On the market side, they talked to several potential partners to fund further product development for each potential indication, although they often asked for exclusive contracts. Across the parallel negotiations, learning was fostered: Some team members served on several teams, and they systematically maintained intranet documentation of the lessons learned, the test results, and the meeting protocols for each cooperation attempt.

Beyond its planned experimentation and residual risk management, the company was twice forced to change its plans in a major unforeseen way. First, they ran into a firm in the United States that worked in the same area but had fewer resources. Each had two working methylation technologies, so they merged (formally, the European firm was the acquirer). Of the four technologies, they picked the best two, one from each company.

The second major deviation from the plan was forced by a failure—they had planned to form many clinical development and marketing partnerships, but these partnerships did not materialize. Therefore, the company ended up in one big partnership with a major pharmaceutical company.

The project was successfully completed in mid-2004 with an IPO and became an ongoing concern. In January 2005, the company reported successfully passing a major clinical development milestone, identifying a biomarker that closely correlates with prostate cancer aggressiveness. This concluded successful market identification for all five initial products in its diagnostic collaboration with the pharmaceutical partner; it triggered an undisclosed milestone payment from the partner.

The Molecular Diagnostics example illustrates how an overall project—"get the startup from technology to proof of concept"—is broken down in pieces, and how the pieces are attacked with a combination of established PRM methods, selectionism, and learning, depending on the uncertainty, complexity, and cost structure of each piece. The example also tells us that it is beneficial to allow the pieces to feed off one another. The parallel partnering trials shared information and lessons that helped each to negotiate better, and learning and modifications were applied to each of the methylation technology candidates as they evolved; indeed, after the merger, a different final set emerged than could have been foreseen.

7.2 Choosing and Combining Selectionism and Learning

As we already discussed in Chapter 6, Japanese consumer electronics companies used a "product churning" strategy in the early 1990s. They introduced scores of trial products into the market, "mutant products such as the refrigerator with the built-in microwave oven, the ambidextrous refrigerator whose doors could be opened from either the left or right, etc."[7] The market determined the winners, and the losers were withdrawn. Many of these products failed, but the survivors covered the market and produced new niches. In this "Darwinian" selection, a trial is launched, the unk unks are revealed, whether they stem from technologies or customer behavior, the successes and failures of the trials are unambiguously observed, and then the "losers" are culled from the "winners."

The Japanese manufacturers could do this ex post selection because they had developed the capability of producing product variants quickly and cheaply, and their customers, at first, did not mind buying products that were subsequently no longer available. However, eventually the Japanese consumer electronics manufacturers were overwhelmed by the cost of launching, and then servicing, so many trial products, and pulled back from the approach in the mid- to late 1990s. Instead, they emphasized selection at the subproject level.

As the Darwinian selection of final products became too expensive, they would, instead, pursue competition within the lab—that is, they would not go through the expense of launching complete product alternatives but would explore and select alternatives before they launched them. This *early* selection was less expensive, but it had one major weakness that the Darwinian selection did not have: incomplete information. In the lab, not all unk unks will reveal themselves like they do in the marketplace. It is our task to explain how the interaction between unk unks and complexity systematically influences the value of the information produced in a way that we can understand. Thus, we can offer a causal framework and build understanding and intuition of how selectionism and learning compare.

Figure 7.1 illustrates four canonical examples of learning and selectionism in projects: instructionalist, Darwinian, sequential, and exploratory.[8] In the traditional PRM box, neither learning nor selectionism is used extensively. This, in fact, is the standard contingency approach to project management; the plan is followed, and preplanned contingencies are implemented as foreseen uncertainties arise. We have already discussed the *Darwinian* example. This is the pure selectionist strategy; projects are run in parallel and allowed to compete, unk unks are revealed, and the best project is chosen ex post. The *sequential* example is the pure learning strategy. Parallel trials are not used, but the project is simply modified over time as unk unks are revealed. The final example, the *exploratory* strategy, combines both learning and selectionism. Parallel trials are conducted at the subproject level, when competition amongst alternatives takes place early, before unk unks have fully emerged, and are thus based on incomplete information. These trials are then incorporated into the overall learning strategy where the project is modified over time.

By comparing these canonical combinations of selectionism and learning, we will consider the differences in the *costs* of applying each approach, and the differences in the *value* of the information that each approach tends to produce.

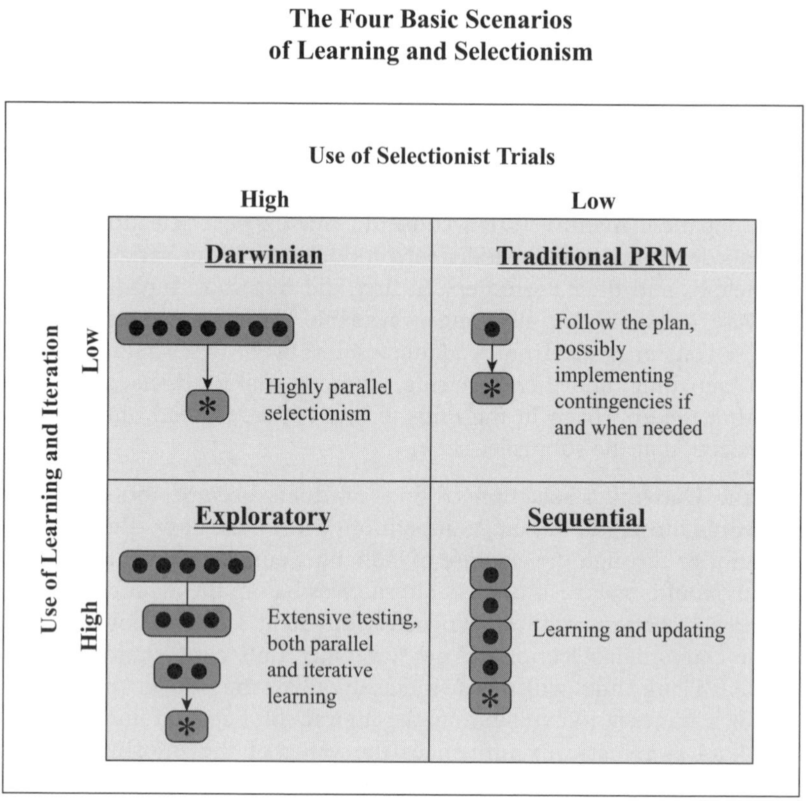

**The Four Basic Scenarios
of Learning and Selectionism**

Figure 7.1 Four canonical examples of learning and selectionism

7.2.1 Understanding the Cost of Darwinian Selectionism and Sequential Learning

We start with the cost side because, with managers, this springs to mind immediately—if parallel trials or repeated iterations are very expensive, they are not affordable. Project managers have intuition about this comparison; indeed, managers sometimes ask us, "How can you possibly afford to do something multiple times in parallel? That's too expensive," not realizing that selectionism can be done on a scale that may well be affordable.

We saw in Darwinian selection that the value of the information is quite good, as the unk unks are revealed in the harsh light of the marketplace, but the costs of launching these products could be quite high. The costs of Darwinian selection include the costs of pursuing multiple solution candidates—which include the cost of personnel and material—and the potential negative impact on the brand. Trials that "do not work"—and typically, most of the solution candidates do not work—may generate negative reactions by the employees and customers involved with the products.

The cost of learning includes activities to identify unk unks, such as experimentation, hiring experts to design experiments, or screening the environment. Also included is the cost of running the iterations themselves, such as testing facilities, personnel and equipment, information leakage to the outside, negative word of mouth if an iteration has a negative result, and the delay from repeated iterations, such as deadlines missed and penalties incurred, a seasonal window that is missed, or the project being outraced by a competitor.

These costs are not to be found in a "universal handbook"; rather, they depend on the organization's capabilities. For organizations that do not have the capabilities of trying out experimental solutions and incorporating feedback from the environment, or that cannot run parallel solution teams, the costs of either approach may well be prohibitive. Moreover, the costs are not precisely known, because the precise nature of unk unks cannot be foreseen. Still, when differences in costs between selectionist trials and learning over time can be estimated, they must play a role in the decision.

7.2.2 Value Comparison of Darwinian Selection and Sequential Learning

In addition to, and separate from, the *costs* of obtaining a problem solution, we must consider the *value* of the solution found. Value refers to the quality or performance of the output achieved by the (sub) project, for example, the quality of the technological solution developed, or the customer response achieved by a new configuration. The value comparison may be even more important than the cost comparison; we discuss it second only because it is less familiar to project managers. They have intuition about costs, but to date, there exists no framework for value comparison.

In this section, we first examine the "pure" case of Darwinian selection, where unk unks are revealed in competition among final project outcomes, the best of which is chosen ex post. We compare this to sequential learning, where a project is changed over time in response to unk unks as they are revealed. Comparing these two relatively pure examples will allow us to set up the more typical situation where selectionism is conducted at the subproject level, where unk unks are not fully revealed.

Clearly, Darwinian selection will be favored whenever parallel trials are cheap and/or delays are expensive (upper left box in Figure 7.1). Learning is minimized within each project, and instead, speed and cost are emphasized. If time is all-important, and the organization has the resources, multiple projects are undertaken so as to increase the likelihood that at least one will be a success. For example, the successful credit card company Capital One uses parallel experimental products. It rapidly develops many new ideas, tries them out in the marketplace, sees what works and what doesn't, backs the winners, and ruthlessly kills off the losers. In this way, it generates more "hits" than its competitors.[9]

If, however, parallel trials are prohibitively expensive or ex post selection is not possible, then sequential learning will be favored (lower right box in Figure 7.1). For example, Internet browser development in the 1990s permitted concept modifications until a short time before launch, enabled by fast prototyping of feature changes. Unk unks were to be expected more on the market than on the technical side—browser use was emerging, and the market was still learning how best to use it. The software kernel and architecture remained stable, but component reuse and quick prototypes allowed the developers to customer-test a new version within a few weeks. Thus, time delay costs were kept low. Parallel trials, in contrast, would have required putting an entire additional team on the project, a very expensive proposition in the face of scarce capacity, and releasing multiple versions of the product onto the market simultaneously would have been confusing to the market.[10]

But what happens when the cost situation is not so obviously in favor of one or the other, or when either Darwinian selection or learning offers a much higher value? In this case, we need to consider the value of information obtained. As we stated previously, in Darwinian selection, the unk unks are revealed and then the best alternative is selected. In this case, we choose the trial based on complete information. But what about learning? So far, we have taken the availability of full information about the emerged unk unks for granted. But this is by no means always the case.

Let us first examine the value of selectionism and learning in the complete information case (ex post selection after unk unks have emerged). Figure 7.2 provides a simple illustration of a project performance landscape, as was introduced in Figure 6.1. The landscape is reduced to *two* influence factors, in order to allow for a three-dimensional graphical representation (the third dimension being the performance, or quality, of a solution).[11]

The left-hand picture shows a simple landscape with only one performance peak, and the right-hand picture shows a complex landscape with many peaks. In projects where complexity is relatively low (see Figure 7.2a), the performance landscape is more easily understood. As unk unks are revealed, we can utilize this new information to change the project plan such that project performance is likely to be improved. In other words, it is relatively easy to chart an optimal path as unk unks emerge. As different influence parameters do not interact, the team can make incremental changes (one parameter at a time), first improving one and then the other. Thus, information revealed early in the project tells us much about what to do later in the project.

Consider, for example, an engineering project where the team knows that it will have to adjust the process recipe as well as fine-tune the composition of the final product to suit emerging, currently unknown, process needs of the client.[12] In a simple landscape, the team can first adjust the process recipe parameters, and when they work well, adjust the final product composition. The second change does not invalidate the recipe choice, as Figure 7.2a shows.

However, in complex projects, new information may or may not be very useful to us. The unk unks, as they are revealed, may be important to us, but the project performance landscape is so complex that we cannot easily use this new information to improve the path that we are currently executing.

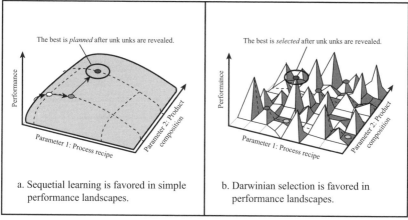

a. Sequetial learning is favored in simple performance landscapes.

b. Darwinian selection is favored in performance landscapes.

Figure 7.2 Learning and selectionism in simple and complex landscapes

In our hypothetical engineering project, complexity means that the process recipe and the composition of the outcome product interact. If the team first chooses the best recipe and then changes the product composition, the recipe now is no longer appropriate and must be changed again. Therefore, there are multiple performance peaks and valleys in Figure 7.2b; the best choice of one parameter changes with the value of the other. In this case, having multiple selectionist trials and then simply choosing the best one ex post, once all the unk unks have been revealed, offers better value than learning and adjustment.

Complex projects favor Darwinian selection, provided the selection can be made with complete information after the unk unks have emerged, while simple performance landscapes will make sequential learning more attractive. Thus, it is not surprising that well-studied engineering design problems tend to follow more sequential iteration, while more complex projects, like the search for new drug candidates, will yield more parallel search.[13]

However, as in the search for new drugs, ex post selection can be prohibitively costly, as it requires completion of multiple projects. In such cases, parallel search often takes place with selection performed early, at the subproject level before the ultimate project performance is revealed, and then the early "winners" are further refined through learning over time. We discuss this case in the following section.

7.2.3 Exploratory Learning and the Value of Partial Information

While the previous section examined the pure case of Darwinian selection, in general, selection of multiple trials may be performed at several stages of a project:

▲ *After design studies*. For example, multiple car concept designs are rendered in CAD models or clay. A concept is chosen based on the "holistic impression" of experts. However, even this holistic impression may not be representative of the ultimate market reaction. BMW found this out when the 7-series introduced in 2002 met a hostile market reaction to the trunk design. This reaction was not predicted after internal and customer tests, yet it forced the company to accelerate the mid-cycle face-lift of the car to 2005.

▲ *After technology tests*. Technology choice is often based on lab tests that cannot incorporate all aspects of the real usage environment.[14] Therefore, the chosen solution may later fail. For example, a tire tread company developed process improvements in a central R&D lab, with the aim of reducing wire breakage in the multiple cold drawing manufacturing stages. It turned out that the result of the technical change depended on the ambient temperature and air composition in the factory, which were not simulated in the R&D lab. This ultimately prompted the company to move technology tests into the factories, despite higher costs.[15]

▲ *After customer or client tests*. In order to predict end-user reactions as accurately as possible, many companies insist on testing project decisions with the client or end customer. But even such checks by the client are often inaccurate. Users' reactions might not be representative of their later behavior under real usage conditions, even when only some aspects of the usage environment are not correctly represented. Thus, client agreements often do not prevent later disputes in complex projects, and market predictions based on consumer feedback are notoriously unreliable.[16]

▲ *After launch*. The opening example of product churning concerns consumer electronics. The success of such products can be diagnosed quickly, within a few weeks after launch in a leading market (such as Tokyo's Akihabara district). This is not the case for products or projects with a longer life. In complex engineering projects or complex consumer durable products (such as cars), success may not be known until a significant part of the product's life cycle has passed, perhaps only several years after launch. Thus, even after launch, the selection decision may yield only partial information. Of course, one could delay selection even further, but that is typically not affordable.

Returning to the pharmaceutical example of the previous section, the cost structure of early lead molecule development in the pharmaceutical industry is low, relative to later stage development. Thus, many lead molecules are produced, and the promising ones are modified—that is, "optimized"—to enhance their chemical reactivity to target binding sites, and thus their potential pharmaceutical potency. This approach is a combination of the two "pure" strategies discussed in the previous section.

Releasing multiple trials of the same drug on the market to reveal unk unks is simply not acceptable in the pharmaceutical industry. If unk unks include death or other undesirable side effects, it is simply unacceptable to have ex post selection after the unk unks have revealed themselves in the death of patients—the costs are simply too high. In this case, the most effective approach is to *combine* selectionism and learning in "test waves" of parallel candidates that are narrowed down early and then optimized over time before project completion (lower left box of Figure 7.1).

The problem, however, with early selection is that it yields only partial information: Not all the unk unks will be revealed at the subproject level. The critical question then becomes "What is the value of this partial information?" Which conditions favor early selection and which do not? Recall that the value of information is measured by the degree to which we can improve project performance based on the information obtained. In the previous section, we saw that the value of early information in sequential learning depended on the complexity of the project. If a project is very complex—that is, the performance landscape is as in Figure 7.2b—early information is less valuable because it is difficult to use it to find the "best" solution a priori. Charting an optimal course is simply too difficult, and in this case, Darwinian selection is favored.

In the case of early selection, where only partial information is revealed, we have the exact opposite scenario: Selectionist trials will *not* be favored in complex projects.[17] This is precisely the most challenging situation for a project team, and it is the situation in which selectionism gets into trouble: If trial selection occurs before unk unks are revealed *and* the project is complex, learning systematically promises a better solution than selectionism.

The reason, briefly, is that making wrong assumptions about an unknown project influence "disturbs" a complex project more than a simple project. Through the many interactions in a complex project, the error in one influence factor has wider repercussions and degrades the quality of the selection choice. To understand this statement in more detail, let us return to the project performance landscapes previously illustrated in Figures 6.1 and 7.2, and now reproduced in Figure 7.3.

In Figure 7.3, in contrast to Figure 7.2, we now consider a situation where one of the influence factors is an unk unk and is not revealed at the time of selection. Coming back to the engineering project example that we discussed along with Figure 7.2, let us suppose that the change in the client's process needs is not foreseeable, either to the client or to the team. In other words, the team takes the process needs as given and does not know that there is an influence that may take on different possible shapes, which have an impact on the project. Rather, the team has an implicit "default" assumption about the client's process needs, an assumption that is unconscious and nonarticulated. This default value is whatever value the parameter takes in early client discussions, although the team is not conscious of it.[18]

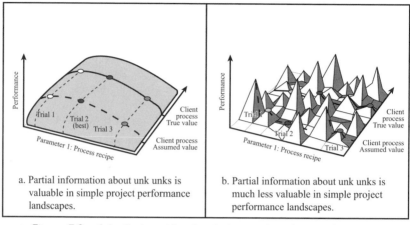

a. Partial information about unk unks is valuable in simple project performance landscapes.

b. Partial information about unk unks is much less valuable in simple project performance landscapes.

Figure 7.3   *Selectionism with unk unks in a simple and a complex project*

We have discussed real examples of this already in this chapter. Take the example of the tire manufacturer who was developing process improvements in a central R&D lab with the aim of reducing wire breakage in the multiple cold drawing manufacturing stage. It turned out that the result of the technical change depended on the ambient temperature and air composition in the factory, which were not simulated in the R&D lab because they were not known to be factors of performance by the team. Similarly, customer preferences in simulated consumption environments might differ from their preferences in actual environments.

Thus, the team's conscious project decisions, in our case the choice among selectionist trials, happen in the "sub-landscape" of the line that corresponds to the default value of the unk unk (in Figure 7.3, this is the dashed line). The trials in the sub-landscape, each corresponding to one decision along the known project influence variable, are marked in Figure 7.3, and one of them is selected as the best candidate.

Now, the problem of selectionism becomes clear. In the simple landscape with only one peak, the parameters do not interact. Therefore, the ranking of the trials is little changed when the true value of the client's process needs emerges (the solid sub-landscape line). The recipe choice is not invalidated by the product composition, and thus, the quality of the selectionist choice is high. In the complex landscape, however, the choice in the default-assumption sub-landscape does not reveal the true best choice. When the true client process needs emerge, the chosen recipe turns out to be inferior (the best trial on the dotted line in Figure 7.3b is one of the worst on the solid line that corresponds to the true client needs). That is, the value of the information yielded from early selection may be of little value in complex projects.

Learning, in contrast, proceeds precisely on the premise that the project keeps evolving, and a new choice is made, after information about the unk unk becomes available. In learning, unk unks are sought out, knowledge gaps are inventoried, and purposeful attempts are made to fill them. In learning, the project team seeks to discover and understand the unk unks. While learning may take more time, it may be more effective than early selection in complex projects.

7.2.4 The PRM Contingency Planning Approach

If both selectionist trials and learning are costly, responding to unk unks becomes very difficult. Both will have to be reduced, although their relative emphasis may remain the same: fewer parallel trials, fewer test waves, and thus, less assurance of a good solution. This means that no sufficiently effective response to unk unks may be achievable, confronting the project with an excessive danger of failure. This has an important implication: If both selectionism and learning are too expensive to be affordable, management should consider changing the project scope in order to avoid unk unks, and thus the need for selectionism and learning. This can be done, for example, by reducing functionality or using proven technologies.

One example of this is the development of the Boeing 777. Boeing developed this plane in response to the Airbus 340, and development time and cost were kept low by using previously proven technology components. Thus, it was possible to develop only one technical solution into a flying prototype; it was simply too expensive to build more than one plane in parallel. In reaction to problems, the prototype was modified, rather than building several models. However, time was very expensive, too, so testing could not go on indefinitely before the company started to earn money. Thus, the first flying prototype was later sold commercially.[19]

The development of the Concorde plane in the 1960s, in contrast, entered much more novel technological terrain. Parallel trials were not affordable (as in the case of the Boeing 777), but because of technical unk unks, the value of time was given less priority than the challenge of getting the plane right. As a result, the schedule slipped by four years due to testing needs, which contributed significantly to a budget overrun from an initial estimate of £135 million to £1.1 billion.[20]

7.2.5 A Combined Choice Framework

We now have at our disposal a logic, a set of criteria by which to choose from among the different approaches for each subproject: The decision should be driven by cost structure and complexity. We summarize these results in Figure 7.4.

When the costs of learning and delay are high, relative to parallel trials, and project complexity is high (upper left-hand box of Figure 7.4), ex post Darwinian selection is favored. Because of the complexity, sequential learning will be lengthy and difficult because causal connections are more ambiguous and harder to identify among the multiple simultaneous interactions. On the other hand, early selectionist trials may not yield any valuable information (see Figure 7.3). Therefore, selection can only be accomplished late.

	Relative Cost	
	Learning and delay more expensive	Parallel trials more expensive
High complexity (many interactions)	Darwinian selection *ex Post* Selection	Sequential learning or reduce complexity
Low complexity (few interactions)	Exploratory *Early* selection and learning	Sequential learning

Figure 7.4 Value comparison of learning and selectionism with complexity and relative cost differences

When project complexity is low and the costs of parallel trials are high (lower right-hand box of Figure 7.4), sequential learning is favored. Here, it is better to pursue a single project and to chart a new course as unk unks are revealed, or, depending on the costs of trials with early selection, to combine selectionism and learning. Early trials are selected and then improved upon over time.

When complexity is low and the costs of learning and delay are high (lower left-hand box of Figure 7.4), a combination of early selection and learning is favored. If time is critical, and depending on the relative cost differences between ex post and early trials, Darwinian selection may be preferred if the organization has the resources to complete parallel projects. As our discussion at the beginning of Section 7.2.3 suggests, late selection (after unk unks have emerged) may be very expensive, as it may require full-scale operation at the client, or a full market introduction in the case of an innovation project. In the case discussed here, complexity is low, so early selection is capable of identifying the best trials.

When complexity is high and the costs of selectionism are high (upper right-hand box of Figure 7.4), sequential learning, although difficult in this case, may be favored. Alternatively, one could attempt to reduce the complexity of the project.

One way to reduce project complexity is by reducing the level of interaction among subprojects, at the level of system components, of project tasks or of relationships (see Chapter 4, Figure 4.4). The more modular a project, the less various subprojects interact with one other, and thus, the less complex the overall project. As we have seen, complexity has a big impact on the value of information obtained, either early information in sequential learning or partial information from early selection. If we can reduce the overall project complexity, it improves our ability to do both learning and selectionism.

It is helpful to classify project interactions within three categories: integrated, sequential, and modular. Figure 7.5 illustrates the design structure matrix for each of these three types. As was discussed in Chapter 4, the design structure matrix indicates the presence of interactions among system components, project tasks, and parties involved. To keep this discussion simple, Figure 7.5 shows only the tasks.

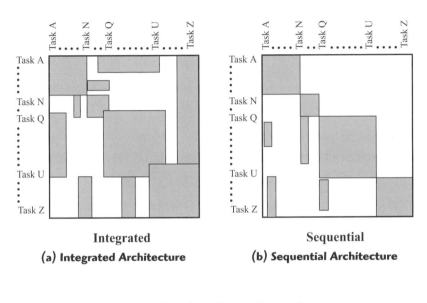

Integrated

(a) Integrated Architecture

Sequential

(b) Sequential Architecture

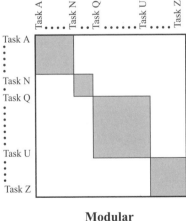

Modular

(c) Modular Architecture

Figure 7.5 Task dependency matrices for integrated, sequential, and modular project architectures

In all three examples (Figure 7.5a–c), tasks A to M are interdependent—they are integrated—and thus are good candidates for a "subproject." We see four such subprojects in all three examples. However, in the integrated case (Figure 7.5a), subprojects are also interdependent. For example, task A depends on task Q, and task Q depends on task A, even though they are in separate subprojects. In the sequential case (Figure 7.5b), subproject dependency only goes one way. For example, task Q depends on task A, but task A does not depend on task Q. In the modular case (Figure 7.5c), subprojects do not interact. They can be optimized separately and carried out in parallel, and unk unks affecting one are not likely to affect the other (unless the project structure itself is unknown).

In an integrated project architecture (Figure 7.5a), subprojects depend on other subprojects in a complicated way. Project performance depends on a complicated interaction of the various characteristics of each subproject. Thus, the overall project can be said to be very complex. In this case, one will have to either do sequential learning or ex post Darwinian selection at the level of the entire project, depending on the relative cost and delay trade-offs.

In a sequential project architecture (Figure 7.5b), because dependency is one way, early selection can be used at the subproject level as long as one proceeds sequentially from one subproject to the other. Thus, a combination of early selection and learning may work well in such an instance.

The "best" case, of course, is a modular project architecture (Figure 7.5c). Here, subprojects do not interact (much), and one is free to conduct early selection on the various subprojects independently of one another. The real challenge here is then to coordinate the various subprojects so that the overall project can be completed on time.

We put this designation of "best" in quotation marks because it represents the view of the project manager. Modularization of a project is often not possible, or only possible by reducing performance of the system that is produced, for example, in terms of size (to make system components separate and noninteracting), in terms of tasks (building buffers into the schedule so that tasks do not happen at the same time and cannot interfere with one another), or in terms of organization (e.g., giving subteams extra resources so they do not need to compete for the same scarce experts or facilities). Because of these trade-offs, determining the system architecture of what is built during the project is usually beyond the scope of the project team. The architecture must be set by senior management in consideration of other projects and business objectives. We will return to the role of senior management in Part IV of the book.

7.3 Reexamining the Circored Project with This New Framework

The Circored project (described in Chapter 2) underestimated the unforeseeable nature of uncertainty and did not compile uncertainty profiles for the subprojects. As a consequence, the project suffered from delays, budget overruns, and damaged careers. How would our selectionism and learning decision framework have helped them?

Let us attempt to apply our decision framework. We do so somewhat speculatively because the relative costs of selectionism and learning were never analyzed; we have to estimate them roughly, in hindsight. Still, the analysis suggests some useful changes in how the project could have been managed.

We define as subprojects the major components of the facility (preheater, bucket elevator and lock hoppers, CFB reactor, SFB reactor, discharge, briquetting), the novel materials, system integration (process control and ramp-up), and marketing (as the product had novel aspects that the customers did not understand). Drawing up uncertainty profiles suggests that the preheater, the bucket elevator and lock hoppers, and the briquetting machines used established technology and were not affected by unknown material properties, so they were not affected by unk unks. We can leave them out of the analysis. Entering the other subprojects, the ones affected by unk unks, into the framework of Figure 7.4 produces Figure 7.6.

Using selectionism for the facility as a whole was simply out of the question, as a planned construction cost of $165 million was too expensive. However, selectionism could well have been applied at the level of some of the ducts and valves made of novel materials. For example, the ducts from composites could have been ordered in two versions, to test and compare them; this would also have allowed the team to react faster when the duct broke and caused a long delay. Similarly, the valve that let hydrogen gas shoot through could have been tried in two different configurations to test for tightness and internal sticking. Double ordering would not have been expensive, as the materials themselves are but a fraction of handling, design, and partner negotiation efforts.

Relative Cost		
	Learning and delay more expensive	Parallel trials more expensive
High complexity (many interactions)	Darwinian *(Ex post selection)* • **Marketing**	Sequential learning/ reduce complexity • **System Integration**
Low complexity (few interactions)	Exploratory *(Early selection and learning)* • **Novel materials**	Sequential learning • **CFB reactor** • **SFB reactor** • **Discharge**

Figure 7.6 Possible learning and selectionism in the Circored subprojects

For the huge reactors and the discharge system, selectionism was, again, too expensive. There was no choice other than to try them out and modify them as the process properties emerged (for example, the walls in the SFB reactor to control retention time). However, once the team had realized that experimentation and learning would be necessary, they could have organized the experimentation very differently: They could have tested the three system components *in parallel*, feeding each with partially processed material. The runs would have been explicitly designed as experiments, not as the production runs that the team attempted almost from the beginning. In this way, the bugs and the verification could have been worked out much faster, with less frustration, and without the need to perform the fundamental verification test of the process in the summer of 2000.

System integration, that is, process control and ramp-up, were highly complex and unpredictable, as the chemical process was new, and hundreds of system parameters interacted in the ore conversion and the properties of the briquetted iron. Moreover, the system behavior could not be simulated beforehand with a sufficient degree of realism (although CAL did develop the process control software, which, after calibration under real operating conditions, was then capable of automated operation). In other words, high-fidelity tests were not available. Moreover, pursuing two parallel process control systems would have been very expensive. Thus, a sequential learning approach was unavoidable. The Lurgi engineers understood this intuitively and approached the ramp-up slowly and deliberately, but this caused some tension and debate with the Cliffs engineers.

Finally, the market introduction of the Circored HBI was also plagued by unk unks because the customers did not understand the product (remember, for example, the fear that residual hydrogen traces would make the briquettes dangerous, or the skepticism over the 2 percent carbon content, which was really a plus for the customers). Again, this was quite complex because customers interacted by word of mouth, and because the HBI entered as an ingredient into a complex steelmaking process. Thus, some experimentation and learning in the customer approach was certainly needed. However, it would have been quite cheap to prepare several advertising leaflets and several "customer briefing story templates" in parallel, trying out on the way that might work best. As the marketing approach was relatively independent of the other activities, a more ex post selection approach would have worked, although in reality, a combination of selectionism and learning would have been needed.

With such an explicit management of unk unks, choosing an appropriate approach, subproject by subproject, the project would have proceeded faster and more successfully. The same unk unks would have affected the team as before, but they would have been handled more quickly, in a more focused way, and with more confidence. Had the facility operated for two years before the 9/11 price decay, it would have already been established as a success, and the historic price drop would not have caused the same fundamental questioning. And quite importantly, the project would have *felt* different to the project team and to the supervising management; all the bad surprises would not have been interpreted, even if unconsciously, as incompetence on the project manager's side, but as normal aspects of building a first-of-a-kind facility.

7.4 Conclusion

We have argued in this chapter that our decision framework of how to choose between selectionism and learning for subprojects can help project managers to plan and execute their project better. In other words, we have proposed a "decision method" that allows the project team to estimate costs and value of selectionism and learning a priori, and then make a better choice of project approach than without the method.

We mentioned several times that project managers may not know the costs of coping with unk unks through learning or selectionism, or the precise value created by these strategies, but that they have a good feel for it. What we noticed is that managers often take decisions to cope with unk unks based on their intuition. How does this square with our proposal of a decision method? An additional use of our framework is to examine the popular rules of thumb, or intuition expressed by project management professionals. Intuition is automated experience that is no longer open to introspection. Intuition is very important for making the entire myriad of snap decisions during the day, for which one has no time to thoroughly deliberate. But intuition is also dangerous because it is based on one's own track record, which may not be appropriate for the problem in hand right now.

The CEO of a venture capital (VC) investment firm, in a conversation with us, gave the following rule of thumb: "If you come to me with a business idea and want money, don't propose selectionism early, in the technology development stage. I expect you to have the technology nailed down. If you tell me you want to try several technologies in parallel, I won't give you a penny. But later, in the marketing stage, I can see selectionism making sense to test out several market approaches in parallel."

We do not want to second-guess an experienced VC who knows what he is talking about. It is still useful to know under what circumstances this rule of thumb would hold, according to our framework. Prescribing learning (and not selectionism) early, at the technology stage, makes sense if the startup (1) is still susceptible to major unk unks, and (2) the situation is complex because technology system elements and market system elements all interact, if nothing is yet determined and frozen. If it is also true that the technology candidates cannot be fully tested for performance, we are in the lower right-hand box of Figure 7.4, and our framework concurs that learning is preferable. On the other hand, if the startup could find a way of testing the technology performance realistically with customers, the answer might well be different.

The second part of the VC's intuition is that selectionism makes more sense during the market introduction stage. Indeed, at that point, complexity and unk unks should be reduced, and so selectionism is more promising (provided, of course, several market approaches can be made at reasonable cost).

Another example of intuition, or a generalization from a small number of examples, is the conclusion in Miller and Lessard's study of major projects, which states that large-scale multiyear projects have no choice other than evolving over time. Again, this is consistent with our framework: Such projects combine major unk unks and complexity, so learning promises higher value than selectionism.[21]

The point of this discussion is not to second-guess project professionals. We are trying to demonstrate that when a project manager enters a novel project, where it is not known what should be expected, it is worth making one's intuition (the principles that one takes for granted, "this is how it's done, of course") explicit, and then to *question the intuition*. Our framework gives the project manager a tool to do that questioning: Where do we have gaps in our knowledge, and thus the potential for unk unks? What do parallel trials cost versus the delay in sequential learning? How high is the complexity of the subprojects and their interactions with one another? Where can we test our solution candidates right away, and where can we not? Can I interpret my intuition in terms of the choice between risk identification and management, selectionism and learning? Does it make sense in the light of the trade-offs?

Our framework gives an operational and conceptually sound rule of how the choices should be made. Part II of this book dealt with the conceptual tools we have in order to understand the conceptual possibilities of dealing with unk unks. Making the choice in a real project team is, of course, a more complicated matter. Solving the managerial challenges of dealing with unk unks in real projects and real surrounding organizations is addressed in Part III.

Endnotes

1. This reflects the current state of knowledge. Although project management experts have discussed selectionism and learning, as we have summarized in Chapters 4 and 5, we are not aware of a framework of how the two approaches compare or combine.

2. Dorothy Leonard-Barton (1995, p. 207) called this *vicarious learning*: "Wait and let the pioneers get the arrows in their backs and learn from their mistakes." For example, IBM, Motorola, and Compaq delayed introductions of PDA (personal digital assistant) products in 1993 in the light of extreme uncertainty about what the market wanted.

3. For further discussion on this, see Watts 2001, and Chesbrough and Socolof 2000.

4. This example is based on Sommer 2004. The name of the company has been changed to protect confidentiality.

5. The process was designed to discover changes between healthy and cancerous cells. Methylation is a natural epigenetic process that occurs when a methyl group binds to one of DNA's four bases, cytosine. The presence of methylation is responsible for controlling the activity of genes by turning them off, like a switch, when not needed. By measuring the differences in the methylation patterns between healthy and diseased tissue, a change in gene activity that could trigger diseases such as cancer is detected. The company had developed an industrial process that was able to read and interpret these methylation patterns.

6. Source: company interview.

7. The story of product churning and its demise is told in Stalk and Webber 1993. The quote is on p. 95.

8. Figure 7.1 is based on Loch, Terwiesch, and Thomke 2001, and on Sommer and Loch 2004.

9. This is reported in Beinhocker 1999.

10. This is described in Iansiti and McCormack 1996.

11. The conclusions are the same if we have many influence variables; for a full discussion, see Sommer and Loch 2004.

12. "Unknown" here means that the needs are currently unknown also to the client, so questioning them won't resolve the uncertainty.

13. See Loch, Terwiesch, and Thomke, 2001.

14. Thomke 2003, pp. 101, 119, refers to the accuracy of tests as *fidelity*.

15. For details, see Lapré et al. 2000, pp. 603 and 607.

16. See, for example, "purchase prediction adjustments" from stated consumer intentions voiced in surveys, reported in Chapter 8 of Ulrich and Eppinger 2004.

17. This discussion is based on Sommer and Loch 2004. A detailed analysis of this situation can be found there.

18. Note that we are even making the optimistic assumption that the team correctly determines the original lab value of the unk unk, here, the client's process needs. If the team does not even listen or conducts the early investigation incorrectly, the assumed unk unk value is, in effect, random.

19. The design of the Boeing 777 did not pose high unforeseen uncertainty; the development was highly complex but fundamentally well understood; see Sabbagh 1996.

20. See Morris and Hough 1987; see also Kharbanda and Stallworthy 1984.

21. See Miller and Lessard 2000.

Putting Selectionism and Learning into Practice

Part II of this book established that the presence of unk unks can be detected, and described how this can be done. We also demonstrated how selectionism and learning are executed in real projects. Part III now develops managerial methods and systems that can help an organization to apply selectionism and learning.

First, we acknowledge that managing the unknown is not something that can be done rationally and in a programmed way, purely by applying methods. It requires a mind-set that allows a project team to cope with the threatening nature of unexpected events that are difficult to interpret. We discuss this mind-set in Chapter 8.

Of course, in addition to the mind-set, the methods *are* important. Chapter 9 presents the essence of the "methods" that we think are relevant. We describe an infrastructure of managerial systems that can be put in place at the beginning of a project to enable the project team to execute selectionism and learning. The systems include the way the organization plans, monitors, and measures progress; coordinates the team members; evaluates performance; and exchanges information.

Chapter 10 focuses on external relationship management of project partners, or external parties involved in the project. External management is of vital importance, and increasingly so as the project becomes larger. It is fundamentally different from team management in that external parties cannot be directly controlled. As we already mentioned in Chapter 3, contractual relationships alone are not sufficient when a project is subject to unknown unknowns. When major unk unks loom on the horizon, a combination of favorable elements is crucial to the success of a project. Contractual bounds must be delineated for a clear understanding across all parties, co-ownership must be established to align incentives, a fair and transparent process must be devised to foster trust, and good relationships must be forged and mutual investments pledged to create commitment.

Finally, Chapter 11 discusses the management of the relationship with stakeholders, or external parties that are not actively involved but who have a vested interest in the project and the power to influence it. This is the "softest" of the management tools, as the project team does not even have contractual means to influence stakeholders, but it is nevertheless important, as stakeholders can sometimes cause a project to fail. Remember Shell's Brent Spar oil storage platform in the early 1990s? Activism by Greenpeace forced a solution that was clearly no better (and by many counts, inferior), purely through influencing. As in Chapter 10, a combination of influencing methods promises most success: explanation (convincing parties that the project supports, or does not impede, their interests), using indirect network ties (drawing on the influence that stakeholders have on one another, to achieve a critical mass of support), and approaching the stakeholders in a way that makes them emotionally positive about the team and the project.

Establishing the Project Mind-Set

Classic project management is about doing your homework up front and then delivering with iron-fisted discipline. As one executive said to us, "We need people who deliver on their promises!" However, discipline, while always important, is not enough in the face of unk unks. Unk unks require a mind-set of not asking where you are in the plan but asking where you are in the first place, and what you really know.

Take as an example Escend, the collaboration software startup company that we discussed in Chapters 4 and 5. Elaine Bailey commented, "It was the mind-set of stopping and asking, 'What do I know and what must I know?' that really helped me." The problem required an open-ended search with an unknown result. It required "switching gears," "putting on a different hat," compared to the execution mode that characterized the other problem areas at Escend (and most of what VCs usually do). This helped her to prioritize the decisive first two weeks on the job. The other problem areas were mere execution. Settling this fundamental question enabled her to make the decision of further investing or shutting down the company. At Option International, the CEO and the company had to remain very open-minded about opportunities that existed in the market, and explore them to see whether they fit within the evolution of the business model, while at the same time avoiding an inefficient deployment of the limited resources of the small organization.

The mind-set cannot be "programmed" in procedures and process, nor should it rest on the shoulders of heroic individuals. In fact, relying on such project heroes could well be damaging for a project, in particular if their "intuition" is not challenged. We see two fundamental sides of the unk unk mind-set, which need to be balanced: first, the openness to look for and see unexpected things, and second, a shared direction, vision, or "map" that maintains cohesion in what the team does. An additional requirement is the team's ability to cope with fundamental changes of the project that take away the "safety" of knowing where one is going. We will discuss these three issues in turn.

8.1 Open-Mindedness: Expecting the Unexpected

Probably one of the most essential differences between a planned approach to project management and the approaches we describe is the state of mind. A planned approach requires a disciplined focus on the plan and the elimination, or at least control, of deviations. A learning and selectionist approach requires, as in the cases of Escend and Option International, a preparedness to be open to unexpected events. This ability to be open to unexpected events has, in turn, three parts: (1) experienced personnel who have seen first hand that projects are, in fact, not always plannable; (2) a culture of encouragement—raising issues rather than suppressing inconvenient observations; and (3) a culture of openness that includes external partners, allowing them to accept the occurrence of unforeseen changes.

8.1.1 Choosing the Project Team with the Right Experience

The Project Manager

Recall the Circored case in Chapter 2, the first-of-a-kind iron ore conversion facility. The project manager was an experienced iron ore mine manager. He was a superb executor, and for him, innovation consisted of introducing new equipment from specialty suppliers, with tight deadlines and budgets. When the team worked its way through the innumerable unexpected problems during the ramp-up, he lost not only his sleep but also all credibility in the eyes of the board because he repeatedly promised success around the corner. He did not have the experience of experimenting and learning in a project. He wanted to make the original plan happen and saw any deviation from it not as a source of information but as an issue to be taken care of. This was costly for the project and for his career.

The Lurgi ramp-up manager, in contrast, was prepared for experimentation and failures, although he had never ramped up such a complex and novel facility himself. He was prepared because he had gone through an "education" by the veteran R&D manager, Martin Hirsch, the father of the technology, who had been involved with the ramp-up of several breakthrough technologies.

A project has to be led by a person. Innovation and project management literature emphasize the crucial role of the project manager.[1] This person needs to have experience with the possibility of unk unks, at least indirectly through others. Moreover, different complexity and uncertainty profiles require different management styles.

In planned projects with variation and foreseen uncertainty, the key ability of the project manager is to be a master planner, an efficient administrator, and a problem solver. He or she must spot deviations from the plan, solve the underlying problems (or have them solved), and expedite the project within the existing framework of the organization.

In the learning approach, the project manager needs to be able to motivate the team to spot upside risks, to turn around experiments very quickly and learn from these experiments. In addition, he or she needs to be able to foster learning in the team (and thus overcome the NIH syndrome). Finally, the ability to mobilize an external network of resources that are not available within the project itself becomes critical, as unk unks may require resources that could not be anticipated and were not provided for. The project manager must also explain the project's mission and sell changes and problems to the rest of the organization. Such project managers have a lot in common with intrapreneurs.

In the selectionist approach, the project manager is an arbitrator among the teams conducting parallel trials. A major challenge is to maintain the motivation of the teams whose solution was not chosen for continuation. To accomplish this, the project manager must be perceived as transparent and fair, and a team player. He or she has also to be a people developer, who can ensure that members of teams that are not selected will feel that they still have a role to play in the organization.

The Project Team

There is a well-established body of knowledge on team composition that suggests how the roles of different team members should complement one another in order to combine to a powerful whole.[2] We do not need to repeat this body of knowledge here.

However, there is a clear difference in team processes that the presence of unk unks requires compared to less uncertain projects. Objectives are less clearly defined and less precisely measurable, and intermediate setbacks are more likely and often more severe. Communication needs to be broader and less structured, and disagreements more prominent, reflecting the need to perform new problem solving, to improvise (see Chapter 5), and to pursue multiple approaches mid-course.[3] In Chapter 9, we will discuss in more detail the managerial infrastructure supporting such less structured team processes. Because of these effects of unk unks, these classical team roles[4] take a slightly different form:

- ▲ *Idea generator.* This refers to synthesizing information from different sources (such as markets, clients, technologies, and processes) to create ideas for opportunities and solutions. When unk unks emerge, this role will be required not only at the beginning (to create good solutions) but repeatedly over the course of the project as new problem solving must be performed out and as multiple design-check-act testing cycles are carried out. Most likely, the role will have to spread over a larger fraction of the project team than in a routine project, where disciplined execution is the main task.

- ▲ *Entrepreneur.* This refers to taking initiatives: recognizing, proposing, pushing, and demonstrating ideas and proposals for solutions. Again, this activity will have to occur more repeatedly and frequently in a project with unk unks and will have to be spread over several team members because opportunities will arise at multiple places in a novel project.

- ▲ *Technical expert.* This refers to mastery of the detailed problem structure and solving the myriad technical subproblems that occur. In large projects, there are, of course, many experts with different areas of specialization. While in a planned routine project the expertise is needed for competent execution, in a novel project it takes on another function: the source of knowledge to create new problem solutions as unk unks emerge. Recall from Chapter 1 how the risk management office put experts on call in order to provide the subteams with problem-solving expertise as residual uncertainty was resolved.

- ▲ *Project manager or leader.* This refers to planning and coordinating the diverse activities in project execution. The manager in a routine project is the planner who then ensures discipline in executing the plan (and not something else). In a novel project, the plan is a fiction, and thus, the project manager is more of an orchestrator as well as an ambassador: It is the project manager who allows new solutions to emerge, and the project content to change, but still ensures the vision's integrity, the common direction of all activities, and the buy-in and involvement of team, partners, and stakeholders.

- ▲ *Gatekeeper.* This refers to collecting and channeling information about important changes in the internal and external environments. The project manager should probably contribute to this, but in large and complex projects, additional access to the various networks is usually required in order to be able to stay up-to-date. The importance of this surveillance function becomes ever more critical as uncertainty and relational complexity of the project increase.

While the project manager needs to put a team together and weld it to a unit, senior management must enable the formation of teams (we will discuss this need in Chapter 12).

In addition to the modification of team roles, the less structured nature of work necessitates three requirements for the profiles of the team members: experience, flexibility, and security. The first, experience, refers not only to deep experience in the technical subject area but also to previously having encountered unk unks and the responses to them, so that the person does not panic or become confused when unk unks emerge. At the very least, the person should be made aware of what unk unks are and the devastating effects they can have (this is the same requirement demanded of the project manager, which we discussed above, in the example of Lurgi's ramp-up manager).

The second requirement, flexibility, refers to personality profiles who are not dependent on fixed routines (as opposed to people who find security only in stable work patterns) and who do not become too attached to work that has been carried out under certain assumptions, so abandoning it does not become too stressful.

The third requirement, security, refers to the right, and need, most people have to be protected from adverse events. In other words, the team member must be assured that he or she will not be personally affected by unforeseen circumstances or project failure, if the team has delivered good work. If project members feel that they are held responsible for overly detailed project targets over which they have no control, they will typically tend to act overcautiously. As one manager put it, "People at gunpoint don't perform better; they freeze."

Can these capabilities be taught? Formal training can definitely help, and we argue throughout this book that a better understanding of selectionism and learning and their trade-off or complementarity will improve the team members' intuition. But such formal training needs to be complemented by on-the-job learning, rotation, and mentoring. The team will be able to cope with unk unks if they have previous experience with them and if they can rely on mentor(s) who, while not necessarily directly implicated in the projects, can allay fears that will undoubtedly rise during the execution of a project that is confronted with unk unks.

8.1.2 Mindfulness: A Culture of Openness

Having a project team that has encountered unk unks before and therefore knows what they are is important but is not enough. If the culture of the organization discourages people from questioning the project's assumptions, they may not dare to initiate responses to unk unks, or they may feel that it is not appropriate because of expectations and social pressures. The mind-set must be "automated" in a *culture*, or *habits*, of never taking things for granted and always looking left and right for things we may have overlooked.

The culture of alertness is well captured in Weick and Sutcliffe's concept of *mindfulness*, which we have already mentioned in Chapter 3 in the context of control-and-fast-response. While this approach is designed to prevent the "system state" from spiraling out of control, the cultural ability to detect and respond to unexpected events is the same as in novel projects. Mindfulness has five components.[5]

1. *Preoccupation with failure.* This means that the organization (in our case, the team) treats any lapse as a symptom that something is wrong with the project plan, something that has been missed and could have severe consequences for execution. Such teams encourage the reporting of errors, they elaborate experiences of "near misses," and they are wary of the potential liabilities of success, including complacency, the temptation to reduce margins of safety, and the drift into automatic processing.

2. *Reluctance to simplify.* It is a common recipe of prioritization to simplify in order to stay focused on a handful of key issues and key indicators. Teams that are alert to unk unks try to simplify less and see more, acknowledging the complex and unpredictable nature of the project. They encourage boundary spanners with diverse experience to challenge assumptions, and negotiating tactics that reconcile differences in opinion without destroying the nuances of knowledge that the different opinions represent.

3. *Sensitivity to operations.* Normal operations, procedures, and processes often reveal observations that have no immediate consequence but are "free lessons" that could signify the development of unexpected events. Take, for example, a conversation with someone from the regulatory agency who says something funny, or the unexpected behavior by a customer in a test, which does not affect the desired test result but could point to a big change in another context. These lessons are visible only if there is a frequent assessment of progress in a multifaceted way, not just tracking indicators. The management team must know in depth what is going on, listen to employees' opinions, and encourage them to express even vague hunches, not so much to react to each hunch, but to be able to detect patterns.

4. *Commitment to resilience.* A key characteristic of unk unks is that no matter how well one prepares, the unexpected *will* happen. Thus, the project team needs the ability to respond to unk unks, to bounce back from disaster, and, perhaps, even to turn them into opportunities. Such teams put a premium on experts with deep experience, skills of recombining bits and pieces of different strategies to a new

whole, and training (this is, of course, closely related to the previous subsection on choosing the project team). Such teams also repeatedly simulate scenarios of surprises and "fire drill" practices of what the team would do. Although the surprise that finally emerges may be different from all the simulations, the team has practiced running through real-time problem solving. For example, the Sydney Olympics preparation included a major subproject of improving wastewater canalization around the harbor to keep debris out. The project faced unforeseen uncertainty because of community relations and unknown ground conditions. The team used a "future perfect" scenario technique of frequently running through the desired project outcomes *as if* looking at them in hindsight and running through the necessary actions to get there.[6]

5. *Deference to expertise.* Diversity of decision making enables an organization to better respond to unexpected situations: faster detection, more knowledge where the decision is made, and more variety in approaches, which increases the chance of finding a good solution. In other words, decision making is pushed down. This does not contradict the requirement to coordinate upward, with the (possibly changing) strategy for the project as a whole, and laterally, with other parts of the project. The prerequisite of delegating decisions is that all team members are informed about the status of the entire project and the key interactions, and that management is well informed about the tasks' status.

Such teams also differentiate between normal times, high-tempo times and emergencies or fundamental changes in the project. Decisions come from the top when progress is normal, but they migrate down when unk unks strike, with collective input to the charting of a new course. The team must know in which mode it operates at any given moment.

These characteristics refer to a team culture that is prepared for unforeseeable uncertainty. Culture refers to shared basic assumptions about the "right" way of operating. This is not something that can be ordered or put in place rapidly; it has to be consistently practiced over a long period of time (on the order of years) before it really sinks in, just as habits or trained routines of a proficient athlete are for an individual.

Instilling a team culture of being alert to unk unks requires the commitment by top management to running projects this way, and then repeated communication and rewards and punishment for behavior in a way that is consistent with the culture. This is not trivial; we often talk to project managers who understand the requirements of unforeseen uncertainty on project management, and even have a corporate statement about it, but report that supervising management falls back into the habit of blaming the team for missing targets when times get tough. When a team receives mixed signals, and supervising management is perceived as capricious and unfair, a culture of responsiveness to unk unks cannot easily develop.

8.1.3 Open-Mindedness among Project Partners

Mindfulness and the willingness to accept unexpected events are particularly difficult to achieve among project partners who come from different organizations. As their project collaboration is temporary, it is harder to establish the trust and alignment of objectives necessary in order not to interpret the unk unk as a manipulation attempt by the other side, and not to opportunistically press an advantage when it arises.

In routine projects, alignment of incentives and actions can be achieved (or at least reasonably attempted) contractually. But this is not possible in novel projects with unforeseeable uncertainty, when changes are major.[7] Contractual arrangements, while indispensable, must be complemented by common expectations, have a shared definition of acceptable behavior and success, and have mutual commitment to win-win actions. Only then are partners usually willing to accept changes without suspicion or counteractions. This can be achieved by irreversible investments in the relationship on *both* sides (that is, both lose when the project suffers), both economically and in terms of personal commitment.[8] We will elaborate on these arrangements when we discuss relationship management in Chapter 10.

8.2 Project Vision, or a "Map" of Unknown Terrain

Openness to changes in the plan, without a common direction, means utter chaos.[9] Openness must be balanced by a flexible, yet cohesive, direction for the project, like a map in unknown terrain. Along such a cohesive direction, team members can orient themselves, coordinate, and censor their own local decisions.

A map of this kind is not only a rational decision coordination device, but it also gives team members a feeling of security in the light of changes that are unexpected and hard to interpret. The following story, illustrating this function of a map, has entered the lore of management teaching:[10]

> The incident happened during military maneuvers in Switzerland. The young lieutenant of a small Hungarian detachment in the Alps sent a reconnaissance unit into the icy wilderness. It began to snow immediately, snowed for two days, and the unit did not return. The lieutenant suffered, fearing that he had dispatched his own people to death. But on the third day, the unit came back. Where had they been? How had they made their way? Yes, they said, we considered ourselves lost and waited for the end. And then one of us found this map in his pocket. That calmed us down. We pitched camp, lasted out the snowstorm, and then with the map we discovered our bearings. And here we are. The lieutenant borrowed this remarkable map and had a good look at it. He discovered to his astonishment that it was not the map of the Alps, but of the Pyrenees.

Thus a map, even a bad one, can help the organization to cope with the threatening and uncontrollable nature of the unknown. It helps to make team members action-oriented and discover other sources of information that may be more appropriate than the bad map. The role of a map in the ability to cope with unk unks, which we call "robust mind-set," is further discussed in Section 8.3.

8.2.1 An Example: Rapid Manufacturing Technologies

What does a map for a highly uncertain project look like? There is no general answer; the structure of the map must depend on the nature of the project and the problems to be solved. It should have a clear representation of the mission of the project, of what really needs to be accomplished, while acknowledging uncertainty and including flexibility in the possible approaches. It may not be detailed, in contrast to a plan, but it must give a sense of direction.

To give at least one example of a map, we refer to the case of the "rapid technologies" center in a major automotive manufacturing company.[11] Rapid prototyping and tooling technologies were originally developed to produce rough prototypes quickly in the development process. The technologies are usually categorized in three groups:

- ▲ Rapid prototyping (RP) is based on the layer-wise generation of physical parts from three-dimensional computer data, like an ink-jet printer that sprays plastic or sinter material, building the part layer upon layer.
- ▲ Rapid tooling (RT) technologies quickly and cheaply produce (stamping, pressing, and molding) tools, either generative (layer-wise) or by molding. While such tools wear out quickly, their low cost and production time allow the cost-effective production of small volumes.
- ▲ Rapid casting (RC) technologies cheaply produce forms for metal casting ("lost" forms, meaning that the forms are destroyed when the metal is poured into them to make the tool). The forms may be produced layer-wise (as in RP) or with sand.

Rapid technologies are predicted to soon move into manufacturing. So-called rapid manufacturing has been propagated as the solution for the vision of low-volume production and customization of cars with short turnaround times. The flexibility obtained by eliminating expensive tooling altogether (RP) or making the tools cheaper (RT) would render the production of lower volumes economically appealing. Vehicles of the future could be produced in small series, and maybe even to customer specifications, at competitive costs.

Experts agree that this vision has real potential for the automotive industry, for some components before 2010, and for mass manufacturing in general by 2015. The rapid technologies center had been charged to develop rapid technologies from a tool used for some prototypes in product development to a tool widely used in manufacturing.

The challenge is that behind the three technology categories above hides a myriad of competing RP and RT technologies, each promising that its view will carry the day. These technologies are being developed by many small technology companies, many of them startups. There is a variety of processes and an ever-expanding list of applicable materials, but all of them still have significant performance limitations. No one knows which of the startups will survive, or which technologies will win. The variety of technologies is too large to be pursued across the full range by one company, even by a large automobile manufacturer. In short, the rapid technologies center has no choice other than to place a few bets, observe the market, and learn as the field evolves. This is a difficult thing to do when one has been charged with achieving a clear objective.

8.2.2 A Roadmap into Unknown Terrain

The project team developed a rough map that is reproduced in Figure 8.1. The map shows three currently available technical approaches, each representing a currently possible combination of different technologies to pursue. Technology sets 1 and 2 are not explained in detail here, for confidentiality reasons. The "process centered" approach means that the new technologies are not emphasized at all, but rather, a dedicated organization produces small series of parts with old technologies, for example, production on flexible turning lathes and mills. This approach can offer speed immediately, but in the long run, it will probably not be able to compete with the new technologies.

The center of Figure 8.1 shows the currently possible, or imaginable, applications of rapid technologies: rapid prototyping (already established), concept cars, individualized cars for specific customers (high-profile or high-paying customers), rare spare parts or parts for vintage cars, and small specialty series (for example, a special edition of 1,000 cars). Requirements are shown to the right of each applications area. On the right-hand side, the figure shows the project vision.

This map allows selectionism, as several of the applications may be pursued in parallel; indeed, the rapid technologies center decided to work on concept cars and one small specialty series. The map also acknowledges that the path from these currently visible applications to mass manufacturing is still unknown, as no one knows how the remaining performance gap can be closed. Thus, learning will have to occur through different application areas pursued over time. This learning will determine which of the applications ultimately leads further.

The vision is sufficiently concrete to give an overall direction while leaving flexibility about how to get there and what performance criteria will ultimately be the most important. Thus, the map can help the project team to manage changes within limits while maintaining continuity for the team to have the feeling that it is making progress and to communicate a cohesive program (rather than a hodgepodge of activities) to the rest of the organization.

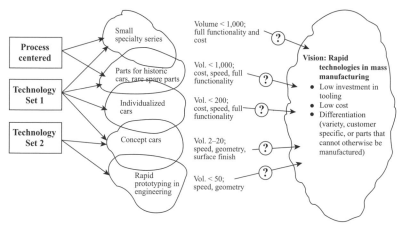

Figure 8.1 A map for a rapid technologies development project

8.3 Robust-Mindedness: The Ability to Cope
8.3.1 Sensemaking and Social Cohesion to Prevent a Team Breakdown

Being open to recognizing unk unks and to changing the project definition is one thing. Being able to cope with fundamental, and possibly repeated, changes in what the project is about is an altogether different, and equally difficult, challenge. This is true for the project leader who makes the plans, but even truer for the team members who may not know the entire background. Fundamental changes in missions and directions can be very daunting and perceived as threatening one's very existence.

Decision making in novel projects (and, in fact, in all ambiguous and complex domains) is heavily permeated by the use of *intuition*. In contrast to analytical and conscious deliberation, intuition is a decision or conclusion that is reached with little apparent effort, and typically without conscious awareness, involving little or no conscious deliberation. Intuition is like "automated expertise" in that it can be learned tacitly (by doing or experiencing) or explicitly (through instruction) and that it is domain-specific (as learning occurs mostly in specific specialized domains, not across the board).[12]

Intuition is used in a much broader set of situations and problems than deliberation. In particular, when situations become complex and ambiguous—that is, open to multiple interpretations, when problems become "wicked" rather than "tame" (see the discussion in Section 3.5 of this book)—explicit and analytical decision methods are insufficient, and people make decisions using at least a heavy dose of intuition. The social psychologist Karl Weick calls this "sensemaking."[13]

We all dislike it when our intuition is violated to a degree that we cannot interpret the situation, do not understand the causal connections, and do not know what the possible outcomes are. It is one thing to like gambling, when we know exactly the possible outcomes and their probabilities (take the numbers, black and white, odd and even in roulette). But unforesee-able uncertainty and ambiguity can be very threatening in all cultures.[14]

In particular, Weick has shown that the breakdown of sensemaking, when the situation violates one's intuition and cannot be successfully interpreted, combined with the loss of social cohesion of the group, can have a traumatic and devastating effect, and even lead to the group's collapse. A famous illustration of such a breakdown is the Mann Gulch disaster, a one-day project in which a team of 16 firefighters was parachuted into the middle of the prairie to extinguish a wildfire at 4:10 P.M. on August 4, 1949. At 6:00 P.M., 12 team members were dead and a 13th lay dying from severe burns.[15]

The firefighter team had a foreman and a second-in-command, and the team was assembled by areas of expertise for one occasion only. In other words, the teams were different each time, and the crew members knew one another only superficially or not at all. Their radio was destroyed during the parachute landing. The fire was classified as a moderate one that could be surrounded and extinguished by the next morning.

The fire did not behave as it was supposed to, given its classification, and produced more noise and heat, but the leaders acted casually. When they walked toward a river, they spread out and lost close contact. In the smoke and the noise, only the leader saw that the fire had jumped the river, and led them up the hill, which confused them. Close to the top of the hill, the leader saw that the fire was racing toward them, lit a fire in front of the group, and ordered them to "drop their tools" and step with him into the fire he had just lit. But the crew did not obey the orders and ran. All of them perished in the fire except two men, who found a crevice through which they climbed down toward the river.

Maclean and Weick's reconstruction of the event suggests the following. The inconsistency between the fire's classification, what the crew members saw, and the behavior of their leaders confused them; they felt unsure in their personal judgment of the degree of danger and the fire's expected behavior. When they spread out in the smoke and noise, the group's identity broke down, and each crew member felt cut off. When a group's cohesion is lost, along with the feeling of mutual commitment and consideration, and of a common purpose, panic sets in. Panic is an intense fear that isolates each individual and sets him out on his own. In panic, the crew members did not really hear or see the leader's action and command, or dismissed it as crazy, and ran for it. The leader survived, lying down on the ashen spot of the fire he had lit, while the wildfire passed him on all sides.

Weick's influential interpretation of the Mann Gulch disaster suggests that a severe disruption of sensemaking, a failure to understand the causality of what is happening around people, together with a weakening of team identity and cohesion, can be so frightening that a team, or an organization, may collapse and be unable to function.

The implications for novel projects are immediate: If team cohesion is fragile because, for instance, stakeholders and various subteams have conflicting interests, and then the project definition changes unforeseeably in ways that people do not understand, or if competing teams in parallel selectionist trials are declared winners and losers in nontransparent ways, motivation may be completely lost and collective action undermined.

8.3.2 A Sensemaking Breakdown in Automotive Development

Consider a case of unexpected changes and a cosmology episode in an automotive development project, in the subproject of the climate control system (CCS). The CCS contains all components and development activities related to the passenger's climate environment, including air ventilation, air filtering, warm-up, and cool-down. One manager explained: "Here, at the air-intake, you find all the problems we have in the development of new vehicles: coordination with other components (e.g., fire-wall, engine) and information release to tooling." Together with the dashboard, the CCS is the subsystem with the most interfaces to other activities.

Due to its many interfaces, the CCS is impacted by complexity over all three of the domains we explained in Chapter 4: component, tasks, and organizational. Therefore, this CCS subproject was subject to many engineering change orders (ECOs) during the last year of development, many driven by changes not in the CCS itself, but in other parts of the car. As a case in point, the engineer responsible for the design of the air intake of the CCS had been constructing a particular component for over a year, based on design assumptions (such as the available space) that were formally written down and "frozen" in previous information exchanges. Subsequently, he had to cope with a total of 18 ECOs, many of which were based on elements beyond his horizon, which thus had no obvious logic. As a result, his sensemaking collapsed, leaving him in severe stress that resulted in his taking extended sick leave.[16]

It is important to note that this breakdown episode, and its destructive effects, were due not to unk unks but to complexity and the associated coordination problems. There were many changes in the design of the car that represented effects of variation and foreseeable uncertainty, nothing beyond traditional PRM. However, as the changes were not communicated or explained throughout the wider development team, some engineers lost their sense of social cohesion ("I am the lowest of the pack, and everyone jerks me around"), and their feeling of understanding what was going on ("How is this possible? This was frozen and settled; is this project in trouble?"). This episode illustrates that complexity, interdependence, and lack of communication can cause inexplicable events for individual team members that are just as devastating as major unk unks.

8.3.3 Robustness: Social Identity and a Map

An example of successfully navigating unforeseen uncertainty and fundamental changes is the effect of the World Trade Center terrorist attack on an investment bank's trading room, and the recovery from it.[17] In this case, we see how both social identity and an overarching vision, a map, allowed a group to survive through threatening changes.

The investment bank, International Securities, operated a large trading room on the 19th to 21st floors of the World Financial Center, directly adjacent to the World Trade Center. The trading rested on a combination of different types of arbitrage, novel interpretations of value linkages of stocks leading to opportunities. Thus, the trading room was one social unit, in which information and strategies were shared, allowing more novel interpretations and trades. Thus, the whole was more than the sum of the parts. Sharing and social cohesion were supported by the manager and personnel policies.

On September 11, 2001, the traders were interrupted by a loud explosion in the next-door building. As they rushed to the window, they saw one tower go up in flames, and a few minutes later, they witnessed the approach of the second plane. A tumultuous escape to the Hudson River followed; when the towers came down, they were all on a boat or already in New Jersey. Fortunately, no members of the trading room were harmed.

Deep uncertainty and anxiety followed, above and beyond the fundamental shattering feeling that the integrity of the United States had been violated. The traders were cut off and did not know whether the Financial Center was also destroyed, whether their jobs would still exist, or even whether the company would survive the disaster. People frantically tried to contact one another, and shared their confusion over a Web site of "accounted-for traders" that the bank put up. After two days, an executive sent a message to everyone: "We are trying to reestablish the systems and contact you," in effect communicating, "Be patient, you are valued employees."

As the chairman of the NYSE announced on September 14 that trading would be back up by the 17th, the bank also decided that they had to start trading again as quickly as possible, and eventually return to the Financial Center. Thus, a project was formulated, although it was not clear how long it would take to resume trading. An important part of the project was to retain the personnel and maintain morale, in spite of terrible work conditions, an undesirable location, and reduced bonuses for that year. It is this part of the project on which we focus here.

By September 17, a makeshift trading room had been reestablished in the basement of a converted warehouse in Escapaway, a small suburban town in New Jersey. To reach the room, one had to weave one's way through several rows of corporate cubicles and corridors. It had no windows, a low ceiling, and walls painted in industrial yellow, and it was filled with desks, phones, laptops, screens, and visibly, American flags.

The first step in their fight to maintain morale was to reemphasize, or reestablish, their identities. The flags boasted, "We are Americans." Then, they taped prominent signs to groups of desks, "20th floor, Equities," "21st floor, Fixed Income," and "19th floor, Risk Management." The desks were rearranged in the same configuration as in Manhattan, and the same people were neighbors. The three floors had been reproduced horizontally. The traders were beginning to feel like traders again.

Then they developed new procedures to operate under rudimentary conditions. System backup and connectivity were greatly decreased, computing power was reduced, and not everyone even had access to a computer. Junior traders had to help by executing trades manually (booking, registering, breaking into parts), an activity that had been performed automatically by the IT systems. This was such a radical return to old practices from a decade earlier that the junior traders did not even know how to do it. The traders resourcefully combined available technologies, personnel, and space to operate. In this process, some traders became clerks, others manual operators, and all shared now-scarce bandwidth in their connection to the NYSE. In the process, they adapted the dress code from business casual to jeans and boots. The changes in roles did not detract from their status as traders; in fact, it is how they reaffirmed their status as traders.

Over time, however, group cohesion was threatened by the drab location—an hour away from the informal networks and information circuit of Manhattan, in a suburb with lots of fast-food restaurants and surrounded by an endless succession of indistinguishable shopping malls. The circumstances began to threaten their identities as sophisticated Manhattanite professionals. One trader remarked, "I have to use a backup chemical toilet; that's unheard of in the securities industry." In December 2001, the merger, options, and convertible bond trading group broke away and established a temporary office in midtown Manhattan. They cited "critical need for access to networks of informal information" as a reason, which obtained them permission from upper management for the move.

It was widely felt that this move could jeopardize the unity of the trading room and, ultimately, the existence of the firm: The move cut the other desks off from vital information that was no longer flowing within the trading room. More deeply, it could mean the end of the trading room, as the manager said, "If it becomes clear that we can trade separately, you wonder what's keeping us together, what's preventing some of us from starting an independent hedge fund . . . This introduces personal economic uncertainty for me, too."

Management's answer to this threat was to emphasize the move as temporary and to present a "map" back to normalcy: a plan of the reestablishment of the World Financial Center room. In the eventuality that the building would be declared unsafe, a backup facility was developed, five minutes by ferry from the Financial Center, in Hoboken. Thus, the firm used selectionism for the return (the Hoboken facility was later turned into a permanent recovery site for future disasters). The map back to normalcy maintained, for several months, two parallel routes.

The race was on—the employees had collectively set a tacit deadline; if the move back from Escapaway did not happen by the time annual bonuses were paid in April (everybody knew the bonuses would be much lower than normal because of the disaster year), people would start defecting.

In the end, the move back took place successfully at the end of March. "Thanks—Welcome Back" read a huge sign next to an American flag over the entrance to the Financial Center as the traders returned. All the desks were once again together. In spite of the six-month displacement, not a single trader had left the organization, and the bank had retained its most precious asset.

Hard work had been invested to maintain social cohesion of the group, and to keep sight of a stable overall direction, in spite of the chaotic and incomprehensible day-to-day changes. Maintaining these two guideposts allowed the organization to maintain morale and continue to function.

8.4 Summary: How to Foster an Unk Unk Mind-Set

In summary, we have seen that the unk unk mind-set requires a culture of mindfulness and the ability of the project team to cope with fundamental changes without losing orientation and morale. The ability to cope is related to maintaining social cohesion of the team and offering a "map" that allows for changes to the project without losing a sense of continuity of purpose and reference.

Mindfulness refers to a culture of constantly looking for symptoms indicating that the project plan may be obsolete—a culture of encouraging problem reports and valuing diversity in problem interpretation. In such a culture, a project cannot be led by simply tracking performance indicators; the interpretation of problem symptoms requires management to engage deeply in what is going on. This includes experience and problem-solving capabilities of the members of the team. The unk unk mind-set is incompatible with a management style of holding people responsible for their targets and milestones without otherwise engaging.

We have seen that teams who lose their frame of reference, their sense of understanding what is going on, become confused when the project fundamentally changes, and may even panic if a sense of common identity and security is not maintained. Team identity refers to the individual team members feeling that they are part of something bigger than themselves, that there is a common purpose and a mutual commitment of the team members. This must be maintained by ongoing social interaction (not only technical information being exchanged) and enforcement of a standard of behavior for the good of the team.

Social cohesion also depends on the ability of the team to understand, and articulate, a common purpose. A "map" is any tool that allows the team to recognize continuity, in spite of project changes, and to articulate what that stable common purpose is.

Endnotes

1. For example, Tampoe and Thurloway 1993, Pinto 2002; in innovation literature, Clark and Fujimoto 1991 emphasize the necessity of heavyweight project managers in their seminal study on the automobile industry, and Wheelwright and Clark discuss the heavyweight project manager in a broader context.

2. See, for example, Boddy 2002, Chapter 8, and Roberts and Fusfeld 1997.

3. See Boddy 2002, Chapter 9.

4. An additional function is sponsorship; we will discuss this function separately in Chapter 12.2.3.

5. The following discussion is based on Weick and Sutcliffe 2001, Chapter 1.

6. This example is based on Pitsis et al. 2003. We will come back to it in Chapter 9.

7. See, for example, Miller and Lessard 2000, Floricel and Miller 2001, or von Branconi and Loch 2004.

8. The importance of such arrangements has been empirically shown by Doz 1996.

9. This has long been known in innovation research; see, for example, two classic references in Quinn 1985, or Van de Ven 1986.

10. Cited from Weick 1995, p. 54.

11. This example is based on Loch et al. 2003, and on an unpublished report to the company that was involved. Figure 8.1 is a disguised version of the map that was produced within the company.

12. A definition of intuition can be found in Hogarth 2001, p. 14; the discussion of emotion is ibid, pp. 62–65, and the comparison to expertise on p. 205.

13. See Weick 1993: "Sensemaking is about contextual rationality. It is built out of vague questions, muddy answers, and negotiated agreements that attempt to reduce confusion, (...) the more frightening feeling that old labels are no longer working" (p. 636).

14. Psychological studies have established that humans dislike unforeseen uncertainty and ambiguity; see, for example, the classic 1985 study by Cohen, Jaffray and Said. The disliking of ambiguity has also been shown in sociological studies across cultures. For example, Hofstede 2001 identifies "uncertainty avoidance" as a measurable tendency (with different relative emphasis) across cultures.

15. The Mann Gulch disaster is recounted in Maclean 1992 and analyzed in Weick 1993.

16. More details of this example are recounted in Terwiesch, Loch, and De Meyer 2002.

17. This example is cited from Beunza and Stark 2005.

9

Putting the Infrastructure in Place: Management Systems

After Chapter 8 has established the "mind-set," or "culture," of managing a novel project with selectionism and learning, Chapter 9 discusses how to manage the project, in terms of systems and methods, which represent an important part of an organization's capabilities. The system offers managers tools that they can use to put selectionism and learning in practice.

9.1 Managerial Systems in Project Risk Management

Managerial systems are necessary for members of an organization to successfully collaborate, so that their efforts complement one another and are aligned to achieve an overall system performance. In novel projects, the environment is too ambiguous and complex for any individual to fully understand. Human rationality is "bounded": Decisions in such environments are too complex for any one individual to comprehend. Project teams cannot "optimize" over the set of all conceivable alternatives, and so management systems are needed to help guide their actions.[1]

Managerial systems consist of routines and decision rules that are analogous to "skills" or "intuition" of individuals. They help in the smooth execution of certain sequences of actions that are automated, and thus executed without apparent effort or conscious decision making. They are also tacit—that is, not articulated: Often, individuals cannot even explain them. Such routines represent stored experience and are effective in the "normal context," in terms of the actions of the surrounding employees and the characteristics of the project environment.[2]

Routines are repetitive patterns of activity in an organization; processes are examples of routines.[3] Routines are de facto organizational memories (analogous to the intuition of an individual)—each organizational member remembering his or her part of activities, and the context of behavior of the actors around him or her, is enough to enable the organization to behave reliably over time, without the need for any individual to fully understand the entire system. Some individual understanding of the entire system is always desirable, but in many cases, it is simply not feasible. Knowledge in organizations is usually distributed and "routinized."

Project teams can perform completely new and original problem solving only in bursts, during short intervals of intense effort by a small group. Original problem solving puts such a level of stress on the team that it cannot be a continuously ongoing process. Original problem solving may result not only in a new solution but in a change of the routines by which the team operates, and thus, in an innovation. Most people find it too stressful to constantly innovate over long periods without the security of routines. This was illustrated by our discussion of the map and sensemaking in Chapter 8, which showed how the map, even the wrong map, provided a sense of routine to the group, allowing them the security to innovate a solution to finding their way back home.

All this has important implications for a project team that must deal with unk unks: The fundamentally nonroutine activity of responding to unk unks must somehow be routinized at a higher level, by learning the (automated) skill of challenging assumptions and experimenting, or of trying out multiple solution approaches in parallel.

Managerial systems are an important component of organizational capability and form the basis of competitive advantage: "Competitive advantage of a firm lies within its managerial and organizational processes . . . the way things are done in the firm, or what might be referred to as its routines, or patterns of current practice and learning."[4]

Managerial systems consist, first, of the formal and informal ways for guiding action and learning. This can happen, for example, through formal problem-solving methods, information collection protocols, such as benchmarking or competitive intelligence, or prototyping guidelines. Managerial systems also provide tools for control through procedures, incentives and rewards, monitoring, and information exchange.[5] Managerial systems are the organization's means of encouraging (or forcing) employees to engage in certain behavior and take certain actions. Due to their routinized and "automated" nature, managerial systems, and the capabilities that they incorporate, may *hinder* a project, if the project has requirements that are incompatible with, or in contradiction to, the organization's managerial systems and routines. As Dorothy Leonard-Barton (1992) noted, "Core capabilities can become core rigidities."

To summarize, a large body of management research has found strong evidence that a firm expects too much of its employees if it simply demands to "solve the problems that come up" and assumes that this can be done from scratch at will. Over time, humans can solve problems only "on the margin"; they need to build upon a foundation of a stable structure of procedures and success criteria. An organization that wants its project teams to successfully deal with unforeseeable uncertainty must give them managerial systems that specify a skeleton of search and decision procedures.

An infrastructure of such management systems must include five areas, which are all heavily affected by the presence of unk unks:

- ▲ *Planning.* What do we plan, and what does a plan look like when it is only stakes in the ground, as opposed to a detailed specification of all foreseen actions?
- ▲ *Monitoring and progress measurement.* What do we measure? Milestones? Achievements? Knowledge obtained? Or actions (such as experiment cycles) executed?
- ▲ *Coordination and relationship management.* How do the various parties in the project adjust their activities with respect to the other parties? In other words, what *actions* should the parties take in response to the information that has been exchanged about their respective status and situation?

▲ *Information management.* Strictly speaking, this is an aspect of the coordination management system. However, it is so important that we want to discuss it in its own right. What is the relevant information that must be communicated across subprojects in order to allow them to adjust to one another?

▲ *Performance evaluation.* What aspects of progress, process, and results should be used to evaluate the performance of the teams and individuals involved in the project? This system is concerned with a mixture of measuring process and output, and a combination of rewards for target fulfillment and upside incentives.

The infrastructure must differ between planned, selectionist, and learning aspects of projects and subprojects. This is briefly summarized in Table 9.1, and we discuss the differences in more detail in the remainder of this chapter.

The instructionalist approach outlined under the "Planned Projects" column of Table 9.1 summarizes the management systems that are well known and established in projects with moderate foreseeable uncertainty. Tasks and targets can be planned, their fulfillment can be monitored, deliverables and structured information is exchanged among parties, and fulfillment is used to evaluate people.

Table 9.1: Infrastructure for Planned, Learning, and Selectionist Projects

	Planned Projects	*Learning Projects*	*Parallel, Selectionist Projects*
Planning systems	· Plan tasks and targets · Work structure and defined responsibilities · Buffer against risk	· Overall vision, intermediate targets · Tasks to learn · Rapid turnaround of experiments to learn	· Collective vision, different roles across projects · Intermediate diagnosis criteria of the potential of an individual project
Monitoring systems	· Target achievement · Progress tracking (e.g., % complete, or deliverables)	· Track "experimentation" · What has been learned? · What problem to solve next?	· Project stopping criteria (relative potential) · Information to be shared among peer projects
Coordination systems	· Fulfillment of deliverables · Coordination via work structure in hierarchy · MBE (management by exception) · Little decision power necessary	· Dynamic and less formal · Long-term trust-based relationships handle changes · Decision power to change approach or targets · Higher problem solving necessary	· Relative progress of the projects · Sharing of learnings · Stopping decisions
Information systems	· Progress, deliverables, actual outcomes of events	· Richer, unstructured information exchange and mutual adjustment	· Overarching over peer projects
Evaluation, incentives	· Target fulfillment · Measurement of output	· Upward incentives on output · "Process quality" incentives	· Shared incentives on output · "Process quality" incentives

Management systems must be adapted in projects (or subprojects) with unk unks. *Planning systems* emphasize milestones of learning, versus delivered results. *Monitoring* must track what has been learned, rather than progress along the planned tasks. *Coordination* must be richer and more flexible, as opposed to a work structure with deliverables. It must include the possibility of rearranging the responsibilities as the character of the project changes. This flexibility must be supported by richer and less structured *information systems*. Finally, the project team should not be *evaluated* on target fulfillment—as the outcome is not under the control of the team, this would cause withdrawal of the best people from learning projects. Rather, upward incentives and process measures should be used.

The right-hand column of Table 9.1 offers a few indications of the differences between planning and selectionist approaches. The overarching idea is that one needs to keep a global view of the various projects while tracking what is being learned in the individual projects. The global tracking allows insights to be shared and individual projects to be stopped when they no longer contribute to the whole. Systems and incentives need to keep the whole of the organization committed, even when the projects to which some of the individuals are allocated have to be stopped.

We illustrate the management systems for learning projects and selectionist projects in Sections 9.2 and 9.3, and then discuss how they can be combined at the overall project level.

9.2 The Management Systems of Learning (Sub) Projects

To discuss the management systems of a learning project, we return to the Escend Technologies example from Chapters 4 and 5. Recall that Elaine Bailey divided the overall challenge of turning around the startup company into pieces, which differed along their uncertainty profiles (Table 4.1). Some pieces could be managed with standard risk management techniques, while she adopted a learning approach on others. The management systems for those parts of the project (customer needs, industry readiness, and product functionality) changed significantly. In this section, we generalize the principles.

9.2.1 Planning System

The traditional planning system of a project with foreseeable uncertainty proceeds according to the spirit of identifying all tasks to be performed, over the entire project, and foreseeing all variants and complications; PRM adds possible risks and the necessary preventive, mitigating, and contingent actions to the plan.

This is not possible in the presence of unk unks, as the tasks may fundamentally and unforeseeably change after an unk unk has emerged. Thus, a rough overall plan (compatible with the sensemaking map of Chapter 8) is necessary that outlines the major steps that promise the potential of

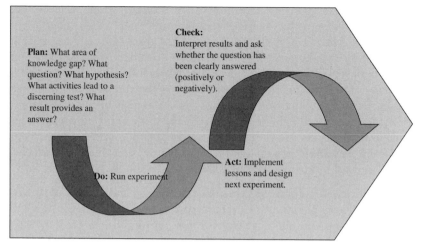

Figure 9.1 Plan the experimental cycle

achieving the project's vision. But this overall plan should be rough, so that it does not become an end in itself. We argued in Chapter 5 that the major vehicle for progress in a learning project is the Plan-Do-Check-Act cycle (Figures 5.4 and 5.6). We reproduce Figure 5.6 here as Figure 9.1, simplified for the purpose of the planning system.

There are several studies available that suggest that effective planning and execution of this basic learning cycle maximizes learning. A detailed plan should be drawn up for the experimental cycle: What is the area of the knowledge gap? What is the hypothesis to be tested and the question to be asked? What are the activities that will illuminate the question? How can we tell whether we have an answer? Once the results are obtained, the next iteration of the learning cycle is planned in detail.

Elaine Bailey understood this. For the three subprojects with unk unks, she did not insist on a plan to the end, but took one question at a time. She planned a set of interviewing trips, an analysis by industry experts, and so on, and then reconvened her management team to reassess what the best next step was. As a result of this systematic learning effort, Escend's search for a functioning business model was very effective.

9.2.2 Monitoring System

The left-hand side of Figure 9.2 illustrates the traditional philosophy of monitoring systems in projects with foreseeable uncertainty. The project manager looks at the planning tool (here, a Gantt chart), draws a line for today's date, and asks whether all activities that are supposed to be completed are indeed complete, and whether the activities in progress are roughly at the required progress status. Admittedly, this view is simplified, but it does capture the basic logic of any tool that compares planned progress to actual progress, perhaps with some variance analysis of causes for deviations.[6]

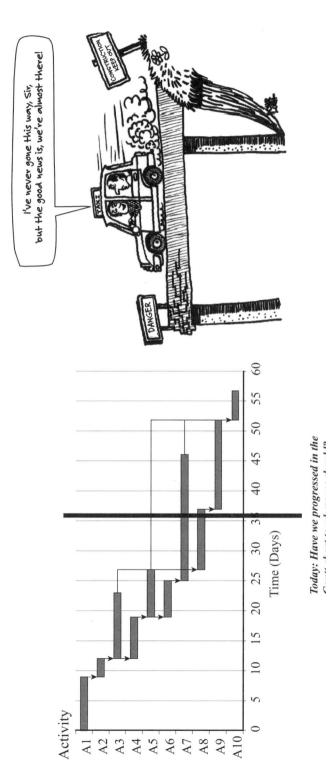

Today: Have we progressed in the
Gantt chart to where we should?

Figure 9.2 Monitoring system assuming foreseeable uncertainty

The cartoon on the right-hand side of Figure 9.2 captures the problems into which such monitoring systems run when unforeseeable uncertainty is present: Progress is often a mirage, and risks and deviations that were not included as "possible" in the risk plan accumulate unchecked (there is no tool to report them!), or are "swept under the rug." Not before the final phase of the project is reached is the team forced to face up to the accumulated problems, at which point often catastrophic deviations, and possibly a breakdown of the relationships within the team and/or with partners, occur. This danger is related to the "double-blindness of planning systems" that we discussed in Sections 3.2.3 and 5.2.2.

We are not saying that progress monitoring should be neglected and that the tools such as the one in Figure 9.2 should be thrown out of the window. These tools are, just like thorough planning and risk assessment, indispensable as the "basic homework." They are often also a good tool to communicate with stakeholders in the project, because they are easy to understand. However, they must be complemented by a monitoring of the knowledge gaps, the open questions, the current hypotheses, and the state of verification or falsification of those hypotheses. This type of knowledge monitoring does not have to be very formal: It may be as simple as the table in Table 9.2 (the table is based on the probing questions document used at Escend, shown in Table 5.2, and develops it one step further as a systematic monitoring tool). But it is essential that the knowledge and hypotheses be explicitly tracked. Being constantly reminded of open knowledge gaps helps the team to remain focused on the large open questions. This can prevent planning and monitoring from becoming an end in itself and double-blind.

It is important that the monitoring document contain causal explanations (the hypotheses as currently available) and reports on changed parameters, as shown in Table 9.2, and not only "output" reports (costs, completion percentage, schedule, functionality achieved, etc.). Studies in complex software development projects have shown that even professional workers are not able to keep track of the causal connections among variables when only outcome information is provided, and tend to make poor decisions and, moreover, oscillate in their decisions.[7]

9.2.3 Coordination and Information System

In a planned project with foreseeable uncertainty, coordination can, at least partially, be defined at the outset, for example, in the form of design rules, interfaces, and schedules that ensure that any party or subproject stays within limits that are compatible with the other subprojects. The information system can then steer interactions during the project by defining clear information deliverables among the parties (for example, completion of activities, shape of certain shared parameters and interfaces, and deviations, or risks occurred).

Table 9.2: Monitoring Table for Areas with Potential Unk Unks at Escend, Status October 2003

Area of Knowledge Gap	Current Hypothesis	Evidence, State of Verification	Next Steps to Close Question
1. Who is the paying customer, and what are the customer needs?	Component manufacturers need design-win tracking, but reps are the gate-keepers and are the major party to be convinced.	Interviews suggest shift of design-win activities back to manufacturers, hypothesis may; be wrong.	Verify information flow and activities across all steps of design supply chain from component manufacturer over rep, OEM, and con-tract manufacturer.
2. Is the industry ready to share data for the sake of better collaboration?	Industry needs collaboration; with clearly demon-strated benefits, companies will share sensitive information.	Still plausible, but design-win seems insufficient as a benefit to get everyone excited.	Verify sales side of relationships.
3. Can the product functionality gaps be closed?	Gaps can be fixed without redesign from scratch.	The technical idea of adding a shell around the kernel (keeping the kernel stable) plugs current gaps, but if sales side becomes critical (see area 2), a new major gap arises.	Test with customers, explore effort of adding sales track-ing module to product.

In learning projects, coordination and communication become more blurred because unk unks make it impossible to define coordination at the outset; coordination has to be repeatedly redefined over the course of the project. Thus, the information and coordination systems become indistinguishable.

As for the planning and monitoring systems, it remains an indispensable "homework" exercise to understand interactions across the various parts of the project (for example, with a classic work breakdown structure, or the design structure matrix that we discussed in Section 4.2) and to establish interfaces. However, information exchanges have to become more frequent and broader, and less confined to initially established criteria.[8]

Project complexity makes it particularly pressing to coordinate and share information quickly. For example, complex engineering projects typ-ically start by cutting the complex problem into pieces. A subteam solves the design of its respective piece, and then the subteams integrate learning about unforeseen system interactions as they emerge. Unfortunately, this type of learning leads to cascading changes and oscillations: If one partner in the project changes, she forces another to change, who, in turn, forces a third subteam to change, which in the end makes it necessary for the first partner to change again. Such oscillations have been widely reported in the engineering management literature.[9]

To avoid such oscillations of system changes, it is important to share information quickly, with as little delay as possible, even if this looks like information overload at first glance. Moreover, engineers should be willing to release preliminary information (even when they are not yet quite sure), which reduces the amount of work done with obsolete information, and to be willing to make small compromises at the component level to make the overall problem solving more stable.[10]

In particular, the information system should have the following characteristics:

▲ *Regular general project reviews* should provide all project team members with an understanding of the status of the overall project and of the subprojects, and of important mutual interdependencies (as they may significantly change when unk unks emerge). A project with unk unks cannot succeed if everyone knows only what is required for his or her job. This includes, in particular, a kickoff workshop (or multiple workshops when the project is large), in which all project team members can familiarize themselves with the project.

▲ Information exchanges must be *more frequent* than in planned projects, and *broader*, which means across multiple information channels. In other words, it is insufficient to exchange information through formal channels, such as change reports; rather, management must provide ample opportunity for members of various subprojects to exchange informal information about any aspect of their work. Only such frequent *and* unstructured information exchange provides safeguards against changes that unexpectedly "ricochet" from one subproject to another. Frequent, unstructured information exchange, although it is inefficient in the short run, helps the project team to be mindful of the unknown (see our discussion in Section 8.1.2).

▲ Information will change over a longer period in a project with unk unks; as unk unks emerge and require a response, a "design freeze" cannot be imposed. The information system must allow project parameters to remain in flux longer, until a judgment has been made that the uncertainty has shifted from unforeseeable to foreseeable. In parallel, the project team members must be kept informed about the uncertainty status (see Chapter 8).

Empirical studies have found evidence that the speed of turning around experiments increases the performance of product development in semi-conductor and software companies. Thomke shows that effective experimentation cycles require not only explicit planning and information exchange, but sometimes also an adjustment of the organizational structure of the project.[11] Figure 9.3 shows the steps of the experimental cycle with an organizational boundary: The design department poses questions and develops the logic of the test. The testing department builds the detailed test and executes it; the results then go back to the design department for analysis and interpretation. Thus, the experimental cycle has two hand-offs, which is almost a guarantee for slow execution. If a project (or subproject) needs to experiment in order to learn about unk unks, the experiments, from posing questions to interpreting findings, should all be concentrated in one responsible hand. At Escend Technologies, this was accomplished by Elaine Bailey, the CEO herself, running the three critical areas and posing the questions. In large-scale projects, the organizational unity of experimental cycles does not happen naturally, so it must be explicitly safeguarded.

9.2.4 Evaluation and Incentive System

We use the term "incentives" here broadly; we do not mean only bonuses or variable pay. We also include more subtle factors that have incentive effects on workers, such as an evaluation that influences subsequent chances of a promotion or a transfer to a more attractive location, the probability of being fired during the next downsizing, or public recognition and peer pressure.

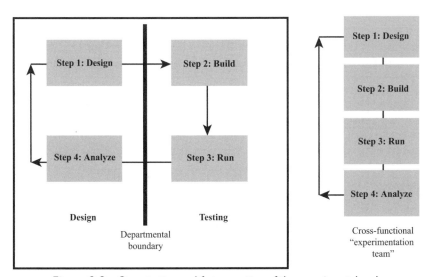

Figure 9.3 Organization and fast execution of the experimental cycle

(Source: Tomke 2003, p. 204)

It is well known from incentive research that productive evaluation needs congruence between responsibility and authority. For example, performance measurement should be supported by job design: If tasks can be separated, the responsibility for them should also be separated, and the domains of responsibility and authority should be congruent.[12] Furthermore, if a manager is responsible for multiple tasks, some of which are uncertain or difficult to measure, no "high-powered" incentives (that means no large bonuses or salary variations driven by the incentive measure) should be used. First, strong incentives would be unfair because they would place the burden of performance variation risk on the individual manager, who is typically risk-averse. Second, strong incentives on the easily measurable tasks would attract the employee's effort (toward the area where his or her effort can be predicted to have a "payoff") away from the difficult-to-measure tasks (which may be as important as the measurable tasks!).

Empirical studies have supported these principles: Performance measurement tends to vary with the character of the work. Thus, output-oriented measures tend to be used for applied projects, while *effort* is measured in risky, long-term technology development, and the issues shift to getting the right people and encouraging breadth of ideas (both from inside and outside) in basic research.[13]

In the context of novel projects, this discussion implies that project performance should be measured as *output* or *process* quality, depending on project uncertainty. For routine projects, the project team can influence outputs, deliverables, or milestones, and can thus be held responsible for them. In a project with unk unks, output measures are not under the control of the team and are thus not appropriate as evaluative measures, except if used as "upward" rewards in the case of success. An upward reward, essentially, amounts to telling a project worker, "This task is so fraught with unk unks that it will have no negative effect for you if it goes wrong; if you *do* manage to make this task a success, we will reward you" (with money or recognition).

However, judging the quality of the effort, process, or method used in a novel project by professional workers is often infeasible, and upward incentives are expensive and can become victims of "private information" and free-riding. Project workers know their tasks better than anyone else; they can hide mistakes and choose not to reveal additional opportunities that would benefit the project but would require more effort from them, or perhaps expose them to a higher risk (for example, of not being able to deliver). Moreover, unk unks are sometimes observable only (or at least long before everyone else) by the front-line project workers. The workers must *volunteer* the reporting of the unk unks (they can always claim later that they did not see it, or saw it later than they really did).

Table 9.3: Process Versus Upward Incentives in Projects with Unk Unks

	Management **Can** Monitor Process	Management **Cannot** Monitor Process
Management *can* observe unk unks as they emerge	Process monitoring, with incentives based on quality of action or method	Adjustment of bonus to problem difficulty & downside protection
Management *cannot* observe unk unks as they emerge		Upward incentives & downside protection

Therefore, the configuration of the reward system must be carefully chosen. Table 9.3 gives a rough guideline of when what type of reward tends to be reasonable.[14] The figure starts with the premise that the manager or team in question faces high novelty and unforeseeable uncertainty (if this is not the case, output-based rewards can be used). The choice addressed by the figure is between process-based evaluation ("has the team diligently and competently used the appropriate methods and processes?") and upward rewards (if the team succeeds against all odds, it is rewarded, while it is protected from negative effects of failure).

If management can indeed effectively monitor the process, the methods used, the diligence of execution, and the level of personal effort of the team members, process incentives should be used. In other words, team members should be evaluated on the quality of the methods, processes, and efforts (left-hand column of Table 9.3), not on the outcome. The outcome is not fully under their control in a subproject with unk unks. Holding them responsible for competent work is the best predictor of success.

Often, of course, the process cannot be effectively monitored, at least not at reasonable cost, because the work is highly professional. Team members have autonomy, they apply tacit knowledge that is difficult to articulate, and it is difficult to tell how hard they work. Sometimes the engineer sitting in his chair with his feet on the table is working the hardest. In this case, it is unavoidable for rewards to be linked to output in some way. However, the presence of unk unks requires a certain downside protection—workers must not be held responsible for unk unks preventing success. People are strongly demotivated by the threat of negative consequences that stem from causes that they cannot control.

The downside protection depends on whether management can observe the emergence of unk unks. If supervising management is so close to the project that unk unks are visible to it as they arise, the performance hurdle can actually be adjusted to the emerging situation: If unk unks made the project harder, workers are given an extra cushion, but if unk unks presented new opportunities that made life easier, or allowed higher performance, the hurdle can be raised.[15] As an example of such hurdle adjustment, consider the following interview given by GE's CEO, Jack Welch, in 1995.[16]

> If I worked for you, you'd say, "I need four!" We'd haggle all day, me making presentations, with 50 charts, saying the right number is two. In the end, we'd settle on three. We'd go home and tell our families that we had a helluva day at the office. And what did we do? We ended up minimizing our activity. We weren't dreaming, reaching. I was trying to get the lowest budget number I could sell you. It's all backward. But if instead you ask people, "give us all you can, give us the best shot at what you can do," then you can't believe the numbers you'll get. You'll get more than you need. There's a trust built that people are going to give their best.
>
> Our plastics business last year had an up year, something like 10% or 11%. But in my view, they had a relatively poor year. They should have been up 30% to 40%. They got caught in a squeeze with prices, and they didn't act fast enough. So, their bonuses were affected. Our aircraft engine business went down $50 million in earnings to $500 million. But we increased their bonus pool 17%. They had a drop, but they knocked the hell out of the competition around the world. They lost $2 billion in sales as the military market and the airline business came down. But they responded to their environment better and faster than their competitors did.
>
> Now, if we were operating under a budget system, plastics would be seen as having a nice year, and aircraft engines would have got a slap in the eye.

This interview is clearly about business units, not projects. But the spirit of Jack Welch's statement is very relevant: Do not evaluate the team on a fixed target. If the environment gets tougher, and one can observe that change, one will adjust the reward hurdle according to the circumstances under which the team was forced to operate. In the terminology of this chapter, we adjust the hurdle to the emerging unk unks.

Finally, true upward incentives should be used only when unavoidable—that is, when management cannot monitor either the process or the emergence of unk unks. In this case, upward incentives provide a motivation to strive for an upside, even if it requires effort and cannot be enforced by management, while the risk of failure is eliminated, which may discourage workers from trying. Again, we include not only monetary incentives but also intangible rewards, such as recognition for achieving something unlikely.

9.3 The Management Systems of Selectionist (Sub) Projects

For the discussion of the management systems of a selectionist approach, we will return to the example of Option International, used in Chapter 6. Recall that the Option International's CEO launched a series of parallel trials over a period of two to three years, in order to redefine the business model of the small entrepreneurial company. In implementing the parallel projects, the organization deployed a lot of good, traditional project management that we would identify with planned projects. For the individual trials, the management systems described in the first column of Table 9.1 applied. But to manage the portfolio of projects, Jan Callewaert had to apply a different management style and implement systems that correspond to what is shown in the right-hand column of Table 9.1.

9.3.1 Planning Systems

One of the main challenges in managing parallel selectionist projects is to keep a global view of all the projects and to prevent them from getting into conflict with one another. Indeed, the overall goal of the parallel subprojects must be a common one. The purpose of a selectionist trial is not to hinder or kill the other trials, but to explore which avenue is the best. If a trial is stopped, the learning that has accumulated in executing it needs to be shared with the other trials.

The essence of the planning exercise is determining the ultimate goal and the collective vision for the portfolio of subprojects (selectionist trials), determining the number of projects one needs to carry out in parallel in order to have sufficient variety, and determining the criteria one will use to diagnose their progress and state.

First, determining and enforcing the ultimate vision was probably the single most important contribution of Jan Callewaert in guiding Option International through its phase of selectionism. He took it on his shoulders to carry through the project of redefining the business model and relentlessly focused the organization on its ultimate goal. While individual teams focused on delivering the best snap-on or mobile phone, he kept in mind, and under control, the overall architecture of what needed to be achieved in the transformation of the business model. He ensured that everybody understood this overall goal, because the survival of the company depended on it. It takes real leadership skills and charisma to render this ultimate goal credible.

The second important task in planning the selectionist approach is to determine the number of parallel trials or projects needed. There is no hard or fast way to determine this, but it has to balance the required variety that needs to be created by the subprojects with the cost of launching additional subprojects. The costs of these subprojects can obviously be reduced, the earlier one can stop trials. The ability to stop less interesting trials early on is thus a very important part of the management systems for selectionist projects.

Third, we already pointed out in Chapter 6 that the ability to stop trials is very important for the success of a selectionist approach. Determining up front sufficiently broad criteria of when to stop projects is an essential part of the planning process for parallel subprojects. In the case of Option International, these criteria were clearly set and evaluated by the CEO, often in collaboration with the board of directors. The evaluation process that led to the end of some subprojects was also enabled by excellent communication between the various teams: The CEO and the project managers were constantly communicating, and they could use the information collected in one subproject to improve the actions taken for another one.

It was also the CEO's strong influence that kept the various subprojects from competing with one another. A selectionist approach is not a war among projects. Success for Option International came through the organization of multiple approaches that communicated with one another, shared learnings, and brought mutual enrichment.

In larger organizations, the important role played by the CEO of Option International needs to be fulfilled by a senior manager, often called program manager, chief project manager, or chief engineer. In the case of Toyota,[17] this person was called the chief engineer and was described as the system architect or the lead designer for the vehicle. He was seen to be the most important technical decision maker on the team. It is noteworthy that this person, in the case of Toyota, had no functional authority over the engineers. The engineers continued reporting to the functional managers. However, the chief engineer was responsible for the vehicle project as a whole—that is, from the early concept stage to the launch of the car and the initial marketing campaign. Most importantly, the chief engineer performed the vital systems integration. While each function was responsible for its subsystem, the chief engineer was responsible for the total vehicle. The chief engineer implements the set-based process, described in Chapter 6, by controlling the process of narrowing the choices and number of sets, by insisting on broad exploration, and by making the decisions on competing alternatives based on the analysis of the trade-offs.

The chief engineer ensured that the overall view for the project was kept, and played a decisive role in closing down the options. This role can be performed in different organizational structures, in functional structures with project managers who have "informal" or "cultural" influence, in matrix structures with a heavyweight project manager, or in dedicated teams. What matters is that someone assumes the responsibility of planning for the selectionist trials, their sharing, and their termination.

9.3.2 Monitoring Systems

As in the learning case and the example of Escend, a Gantt chart is not sufficient to help us monitor parallel projects. Monitoring in this case will require the constant evaluation of the relative contribution of each subproject to the ultimate goal of the project. How much closer does an individual project bring us to the end goal? In the case of Option International, the overarching question was not whether the subprojects were on time (though this was an important challenge at the level of the individual trials). The key question was to what extent the individual projects had the potential to bring the organization closer to a viable and sustainable business model.

Comparing the relative progress of selectionist trials is not easy, and can even be dangerous to the overall project objective. In particular, subprojects do not move linearly at a similar pace. As we discussed previously, selectionist trials are often used for complex (sub) projects. It is difficult to track linear progress in complex projects, because just when you think one team may be about to discover the solution, an unk unk hits and changes one aspect of the project, which then changes others, and so on. Complex projects are highly nonlinear systems, and as such, they are notoriously difficult to track.

In addition, if subproject teams know that their progress is being compared directly to that of other teams, the selectionist approach will become like an America's Cup race: The objective will become, either consciously or subconsciously, to beat the other team. In America's Cup racing, the objective is not to achieve a fast time, but to block the other team, and great effort, and time, can be put into this objective.

Thus, selectionist trials should only be stopped when it becomes clear that they will not find an acceptable solution. This is not easy for organizations, as these trials can be relatively expensive, and it is tempting to shut a subproject down when it appears to have fallen behind. But in complex projects, impressions can be misleading. Monitoring of the selectionist trials should thus be based on the criteria of the project objective, not criteria relative to other selectionist trials. This is difficult for organizations to implement, but it is at the heart of selectionism.

While projects should not be "compared" to one another, they should learn from one another. Here, monitoring can help a great deal. The global project manager needs to monitor how knowledge (both tacit and codified), developed throughout the collection of subprojects, is deployed rapidly and effectively. However, caution is needed in transferring learnings from one project to another. In complex projects with unk unks, one should avoid taking the best aspects of one project and incorporating them into another. In complex landscapes, subtle differences between two projects can render these best aspects useless, or worse (see the discussion in Chapter 7). What is important is sharing the overall learnings across projects. What have we learned in one project that helps us to understand the landscape of another project?

9.3.3 Coordination and Information Systems

The fundamental difference between selectionist trials and experimental learning is the relative autonomy and independence of the selectionist trials. Selectionism is about testing alternative approaches to the (sub) project objective. Thus, selectionist trials should be given the time and freedom to demonstrate whether they will be effective in meeting these objectives or not. Thus, coordination in selectionism takes a very particular form: We need to keep the selectionist trials relatively autonomous while allowing for communication across projects to leverage key learnings.

In addition, we may need to stop some trials when it becomes clear that they will not yield satisfactory solutions. The selectionist trials, then, should not be so autonomous as to create "winners" and "losers" of the various project teams. Coordination in selectionist trials requires clearly specified "fair process" from the beginning.[18] All players must fully understand the game that is to be played. When venture capitalists fund multiple startup ventures in the same "space," the startups are fully aware that if the market place rejects their particular approach, the VC is free to walk away without offering any solace or safety net to the founders. The VC is the hub and the various selectionist trials (startups) are treated fairly autonomously.

In project organizations, the rules of the game are likely to be different, and thus, coordination systems are likely to differ as well. The "winners" and "losers" are likely to be from the same organization and are likely to continue their working relationship regardless of the outcome of the selectionist trials. It is perhaps quite easy to determine up front the criteria for ending subprojects. However, experienced project managers know that the reality of stopping subprojects is not so simple. From the general experience of how to stop failing projects, we know that this is a difficult task for many reasons.[19]

- ▲ Often, the teams and the organization do not want to accept that the subproject is not leading anywhere; teams suffer from the collective belief that the project is going in the right direction and turn a blind eye to negative information.

- ▲ Individual managers do not want to admit that their project is not delivering the optimal results, because it does not fit the image they have of themselves as good leaders.

- ▲ Even if subprojects encounter problems, the organization rarely wants to view this as a negative warning signal because novel projects are supposed to run into risks, and it is the role of a good project manager to overcome these risks. Finally, organizational bureaucracy often makes it difficult to shut projects down.

The same mechanisms can apply with subprojects in the selectionist approach if the subprojects are given too much autonomy. Thus, there needs to be sufficient coordination among the subprojects so that those subprojects that are halted should not be considered as failures, but rather as elements of the final solution. Specifically, one should

- ▲ Establish at the outset of the project clear "rules of the game," noting that many of the subprojects will have to be halted.

- ▲ Ensure that stopping a project has no negative effects on the teams and the individuals leading the subprojects.

- ▲ Use outside resources to help evaluate the progress made by the subprojects.

- ▲ Avoid all the team members of the subprojects being cheerleaders, but by all means, have a few critical members on each team.

▲ Have a formal review process with adequate time set aside.

▲ Keep a high level of common identity among the teams.

▲ Ensure that the whole project reminds itself constantly of the ultimate goal to be achieved by the selectionist approach.

Option International guaranteed that failing projects were stopped appropriately by constantly evaluating the contribution the projects made to the overall goal of the redefinition of the business model. The CEO used the board of directors as a sparring partner in the evaluation of the progress of the subprojects. In addition, the need to report to financial analysts on a quarterly basis gave an almost natural rhythm to the process of evaluation.

Thus, organizations are faced with an important challenge in the coordination of subprojects in the selectionist approach: How much autonomy do you give the managers of the subprojects, both in terms of the goals to be achieved by the subproject (goal autonomy) and how to go about achieving these goals (supervision autonomy)?

The degree of supervision autonomy has to do with the way in which management exercises oversight through the specification and supervision of operational activities. A project group with greater supervision autonomy has greater local discretion, permitting greater heterogeneity in day-to-day activities. Greater supervision autonomy allows for innovation in problem solving and provides an inducement for the team members to exercise greater individual discretion, leading to greater motivation and commitment. Supervision autonomy also helps to minimize the strain on the information-processing capacity of the organization.

The whole point of selectionist trials is to try different approaches, so project teams will need a sufficient level of supervision autonomy to remain independent from other selectionist trials. However, the autonomy of the trial teams must be constrained in that each team while remaining independent, must not have the freedom to mimic what might be perceived as the current "best practice" among the many selectionist trials. The America's Cup metaphor was mentioned in the previous section: One cannot allow selectionist trials to "react" to the actions of other selectionist trials in an attempt to stay ahead of the pack. Each team must fully explore the path that it has been dealt, and not hop on to another's path that the team might see as more promising at the moment. Only then can the complex solution space be properly explored.

The degree of goal autonomy has to do with the way performance goals are set. At one extreme, managers may allow a team complete latitude in terms of goals, focusing on possibilities and opportunities. At the other extreme, managers may be directive, defining very specific goals and outcome criteria. Traditional project management will argue that clarity, measurability, and unambiguity in the goals of the project are key factors for success. The value of clear authority structures and working relationships for a project is seldom questioned. Project management preaches that goal autonomy should be relatively low in order to perform well.

Selectionist trials also need clear goals, as these goals will be used to "select" the trial, or trials, that will continue. Without clearly stated goals, one cannot have a fair process by which some projects are stopped and others continued. These goals must be specific enough for all to agree as to whether one project is meeting these goals better than another. However, one must take great care not to confuse the goal of the project with the "how" of the project. Too often in project management, the project plan comes to be seen as the goal in itself, rather than a means to achieve a goal. Goals for selectionist trials must remain high-level enough to grant each trial the supervision authority it needs to explore the solution space.

In the case of Option International, the combination of both goal and supervision autonomy was realized by stimulating the entrepreneurial behavior of the subteams. The CEO, being the prototypical example of a dynamic entrepreneur in the high-tech business, set the standard for the rest of the organization: He demanded a high level of entrepreneurial spirit from his collaborators, and those who could not cope with it gradually left the organization.

9.3.4 Evaluation and Incentive Systems

The argument we developed about incentives for learning projects also applies to selectionist projects: In the context of high uncertainty, it is difficult to link incentives to output alone. The incentives need to be linked to the process quality and, as was illustrated with the quote from Jack Welch, must be stimulating to get the best out of the people, after adjustment for the level of the emerging unk unks.

In parallel selectionist trial projects, the incentives must additionally reflect a clear message that the "failure" (that is, the stopping) of one subproject cannot be the yardstick for evaluation. Incentives have to create a common commitment to the ultimate goal. But they also need to ensure that the quality with which the individual subprojects are implemented is of high standard, and that information is readily shared across the project.

One way to support this somewhat contradictory set of incentives is to create an "expedition effect." Imagine an expedition with several ships or teams. They know that their best chance of succeeding is to cooperate. Yet each of the ships or teams needs to perform at its best in order to make a real contribution to the expedition and not slow down. And they also know that their best defense against unknown dangers is to constantly communicate with one another. Expeditions succeed when the members of the team know they will all share equitably in the end result. They also cooperate because of peer pressure and because it is in their best interest to survive during the expedition. It is this feeling of being part of an expedition that one has to stimulate through the evaluation and incentive system.

A "financial incentives" approach to creating an expedition effect would contain   incentives based first on the end result, such as stock options or a significant bonus that is determined primarily on the overall performance— that is, the result achieved by all parallel projects collectively—and second on the process quality of the individual trial projects.

A concern often expressed about purely group-oriented incentives is that they dull individual effort and individual stretch for creativity.[20] Incentives experts have, therefore, proposed a "win and audit" approach: Incentives *do* contain a bonus for "winning," for producing the trial project on which the final solution is based. However, if one team's trial is chosen, the bonus is not yet earned; rather, winning triggers an "audit" that examines whether the team has shared information and collaborated with the other teams (this can be done, for example, by a peer review). Only if the team has behaved collegially does it get the extra bonus.[21] However, incentives that reward individual teams, and thus cause the teams to compete, have to be viewed with caution. Much evidence shows that competition suppresses collaboration and may even push some employees to the extreme of changing their work, not to improve results but to prevent others from winning.[22]

Finally, the above outlined financial incentives approach is insufficient and may even be counterproductive. Financial rewards can bring about certain specific actions, but they are weak in producing consistently collaborative and constructive behavior whenever the employees have discretion and autonomy in their work and cannot be fully monitored.[23] Needless to say, this is the situation novel projects face.

Actual day-to-day behavior in parallel teams is driven by social interactions: status, relationships and group identity, and role models. The first concerns status and recognition, both by peers and by management. Recognition is an intrinsic need that people have everywhere (although its expression is culturally specific to countries and organizations).[24] Winning itself carries status, even without any emphasis placed on it by management. This pushes teams into competition. If, however, management consistently recognizes and acknowledges sharing and collaboration efforts, this will, over time, also carry status and counteract competition.

Second, personal relationships across teams, encouraged by shared events (especially important at the outset), and repeated emphasis on the shared endeavor and common goal of providing the best outcome for the project overall also serve to emphasize a common group identity.

Third, the behavior of the team leaders sets the tone. If the leaders are competing and winning types, team members will take the clues and emphasize competition rather than sharing and commonly supported trial selection. If the team leaders get to that position by collaboration, the tone of the teams will be more collaborative. Personnel selection (see Chapter 8) is critical in determining the team's character and working mode.

9.4 Integrating Learning and Selectionist Pieces into the Overall Project

In the first three sections of this chapter, we discussed the management systems that support learning and selectionism in one subproject. However, we saw in Chapter 4 that a large project is not usually afflicted by unk unks in all its pieces. Every project has pieces that are well understood and can be managed without selectionism and learning. For example, the Escend Technologies project from Chapter 4 had three areas with potential unk unks and seven areas where it was clear what had to be done. And the case of Option International is a combination of a selectionist approach for the development of the business model and a learning approach for the technology development. How can these different types of subprojects be integrated at the level of the overall project?

In this section, we offer three principles: In the overall project plan, the areas (subprojects) threatened by unk unks need large buffers; in the spirit of coordination in concurrent engineering, the other subprojects should understand what deliverables and information they need from these subprojects in order to start their own work; and finally, the subprojects with unk unks need to provide uncertainty status updates and "go" signals to the other subprojects that depend on them. We explain each in turn.

9.4.1 Buffers for the Subprojects Threatened by Unk Unks

The activity areas (subprojects) that are subject to unk unks must be managed with experimental iterations. This implies directly that the duration and budgets of these areas cannot be precisely predicted or planned. Therefore, they must be given a *large buffer* that explicitly incorporates this lack of knowledge into the project plan. Figure 9.4 shows an illustrative high-level turnaround project "plan" at Escend, at the level of the subprojects from Table 4.1. From the diagnosis, we know that the first three areas were judged vulnerable to unk unks.

Areas 4 through 9 were straightforward; Elaine Bailey knew what had to be done, and it was possible to swiftly execute these areas. For the first three areas, however, no one knew how long it might take in order to get a clear picture and to understand what needed to be done. Elaine had to go on a learning mission. Large buffers (in white) in the Gantt chart graphically express this knowledge lacuna. In addition, the plan is incomplete; that is why we set the word in quotation marks. While the immediate actions to "stop the bleeding" are known, the really important actions of tapping the potential market (if it exists) cannot yet even be written down in a plan. Figure 9.4 contains a "ghost" activity whose content is concealed. In some sense, the Gantt chart in Figure 9.4 does not contain a lot of information, but sometimes it is important to clearly illustrate how *little* one knows.

9.4.2 Clarify Dependence of Other Subprojects

Given that the subprojects with unk unks have unpredictable completion times and will produce information that we cannot yet describe, the other subprojects will be affected and cannot be planned either, neither in their timing nor in their content. It is therefore important for the project team to understand which subprojects are susceptible to unk unks. In a way, a high-level design structure matrix (DSM, see Chapter 4) should be drawn up just to see which subprojects are "immune" to effects emanating from unk unks.

For example, at Escend (Figure 9.4), subprojects 4 through 9 are pretty much independent of the unk unks arising from market status and product functionality. These subprojects are concerned with stopping the bleeding from the current company situation. They represent a defensive move that must be completed anyway, independent of the findings in the first three subprojects. Therefore, areas 4 through 9 should proceed as quickly as possible. The real product development and market approach, however (the "ghost" activity in Figure 9.4), cannot even be defined (not to mention started) before the unk unks have been resolved and substantial information is available about the shape that the market is taking.

9.4.3 Transfer Preliminary Information

In order for other subprojects to start and to progress, the status of the unk unks, and the subprojects affected by them, must be communicated. The uncertainty status can be illustrated by *preliminary information*.[25] Imagine building a house: You cannot afford to delay the kitchen planning until you have put up the walls. But if you start the kitchen planning too early, using preliminary floor plans from the architect, you are likely to do it twice. You need a new way of exchanging information between the architect and the kitchen planner. Currently, your kitchen planner's idea of concurrent engineering is that he should receive the floor plans, as he did in the past, just six months earlier. He doesn't understand that the nature of the information has changed.[26]

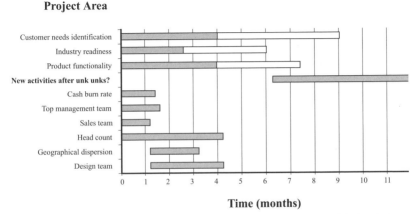

Figure 9.4 High-level "plan" for the turnaround project at Escend Technologies

The uncertain status of information that is exchanged across subprojects can be made explicit by labeling the information in terms of *stability* and *precision*. For a given amount of knowledge, information precision and information stability may be in conflict with each other, as the following everyday scenario illustrates.[27]

A traveler flying from Philadelphia to Paris wants a friend pick her up at the Paris airport. The day before the trip, the arrival time in Paris is uncertain. Thus, any information forwarded to the friend will be preliminary in nature. The traveler can ignore this uncertainty and communicate an arrival time of 14:34, which is precise information but unlikely to be stable. Alternatively, she can focus on information stability and say that she will arrive between 12:00 and 18:00. As the journey unfolds (e.g., before boarding, after take-off, at the baggage claim), the uncertainty of the arrival time is reduced and the preliminary information is revised repeatedly, until it is fully stable and precise (as she leaves the airport).

Note that the two strategies each impose different costs on the friend: The precise information may force the friend to change her own plans at short notice if the plane arrives later or earlier than planned. The stable strategy, on the other hand, forces the friend to keep her diary open for the entire afternoon, blocking any other appointments.

Initially, little information on the resolution of the design decision is available (low level of knowledge), and information is neither stable nor precise. As problem solving progresses and the level of knowledge increases, information is repeatedly communicated with changing levels of precision and stability. At the end of the problem-solving process (high knowledge level), the design solution is in place. Now, information is both stable and precise.

Figure 9.5 shows two strategies of dealing with preliminary information.[28] An iterative strategy emphasizes precision: It utilizes the information in every detail, but as the information is not stable, the response may have to be changed (which can be expensive if rework results—that is, if the response has first to be undone and then reperformed). A set-based strategy emphasizes stability over precision: It uses only "ranges" of possible outcomes, which avoids rework, but the other subproject may not be able to do useful work based on the imprecise information. When there is no unforeseen uncertainty and ranges of outcomes and their probabilities are known, a "best" combination of precision and stability can be chosen, considering probabilities and relative costs of rework versus having to wait.[29]

This definition of preliminary information, stability, and precision is, of course, based on foreseeable uncertainty, or in other words, on a situation where all possible outcomes can be described beforehand. This is not possible when we have unk unks. But the same spirit of the use of information applies to an information exchange subject to unk unks, although the form of the preliminary information cannot be "optimized."

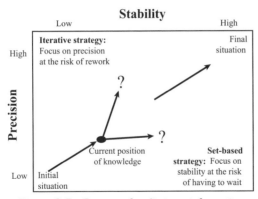

Figure 9.5 Formats of preliminary information

For example, let's again consider Elaine Bailey's situation at Escend Technologies. When it became clear in late 2003 that design-win tracking was not the only important customer benefit and manufacturer reps were not the most important target group, this had implications for the product redesign. Hypotheses arose about the "ghost activity" in Figure 9.4, preliminary information concerning the dependency between the market emergence and the product design: Escend may need to add input and reporting capabilities for manufacturers and OEMs, or ordering capabilities for the contract manufacturers, or distribution capability to enter sales data of the final product. An iterative strategy would imply that the design team work on (or at least prepare) each of these possibilities, discarding the work for the option that was finally chosen. This had the advantage of maximum responsiveness, but it was very expensive, beyond Escend's resources. A set-based strategy implied doing only preparatory work allowing *any* design change to be made, and otherwise waiting. Escend waited until Elaine felt confident, and then took a bet on the distribution capability.

The key lesson is that the overall project needs an information exchange strategy across subprojects that explicitly acknowledges the uncertainty and vulnerability to unforeseen changes. A simple tool that embodies Figure 9.5 can serve as a management system, sufficient to capture the key uncertainties. Overformalization should be avoided because formalized tools easily deteriorate into "double-blindness," claiming precision and knowledge that are not really available. The key questions that should be answered are as follows:

- ▲ What information from the unk-unk-fraught subproject does the other subproject need? (This should come out of the dependency DSM; see Section 9.4.2.)
- ▲ What does the information-delivering subproject team know? Where are the limits? Can they deliver partial information, or a range of possibilities, that are stable? Or is the status still in flux in unforeseeable ways?

▲ Does the dependent subproject have enough information to commit to a certain course of action? Is it feasible and affordable to start a course of action and later change it (iterate)? Or should the team keep multiple courses of action open, or simply wait until more information is available? The answer depends on relative costs and urgency.

▲ Update the status of information at learning points and communicate the updates to the dependent subprojects. If both sides understand the dependency, information can be exchanged in a targeted way, which helps to reduce information overload. The two subproject managers can decide together when enough information is available for the dependent subproject to start at full speed.

The benefit of this management system is not in planning, or optimizing, information exchange. The benefit lies in *making explicit* to the subprojects how they are dependent on one another, and where the dependencies are affected by emerging unk unks. This makes the teams aware that they have to adapt to one another in unforeseeable ways that emerge during the project, and helps them understand where changes come from. The key benefit, in other words, is not in planning but in sensemaking and mindfulness of mutual adaptation (see the project mind-set in Chapter 8).

The precise nature of the vulnerabilities to unk unks and of the dependencies across the subprojects is so specific to different projects that no general statements can be made. However, the need for the teams to understand the unpredictable nature of the subproject interactions, and to deal with them explicitly, is universal in novel projects. The *simplest* tools possible should be used to capture this insight.

Endnotes

1. The term "bounded rationality" was first coined by Simon (1955).

2. See Nelson and Winter 1982, p. 73; see also the discussion of intuition in Chapter 8.

3. Nelson and Winter 1982, p. 97 and pp. 99–100.

4. Teece, Pisano, and Shuen 1997, p. 518. Leonard-Barton (1992) referred to the "position" as the "technical systems."

5. See Leonard-Barton 1992, pp. 113–114.

6. The Gantt chart in Figure 9-2 is based on the unmanned aerial vehicle (UAV) project from Chapter 3.2.1; it corresponds to the network flow diagram in Figure 3.2. For an example of a more complex monitoring system, see Pillai and Rao 1996. The tool presented there is more complex than what we show in Figure 9-2 in the sense that schedule, budget, and "progress" are combined, with the possibility of cross-comparisons and, therefore, causal risk monitoring and analysis. However, the basic philosophy of comparing actual with planned progress is the same.

7. See, for example, Sengupta and Abdel-Hamid 1993.

8. This is already discussed, for example, by Shenhar and Dvir 1996, p. 618.

9. See, for example, Allen 1966, Mihm and Loch 2005.

10. See Cusumano and Selby 1995, Chapter 5, and Mihm and Loch 2005.

11. See Iansiti and MacCormack 1996, MacCormack et al. 2001, West 2000.

12. An influential article, Holmström and Milgrom 1991, established this principle in a formal model that has become widely accepted.

13. See, for example, an overview in Hauser 1998; see also Loch and Tapper 2002.

14. Table 9-3 is based on Sommer and Loch 2005.

15. Subject to respect for transparency and fair process. Fair process is discussed in Chapter 10.

16. Cited from Loeb 1995.

17. See Sobek et al. 1999.

18. We will discuss fair process in detail in Chapter 10, when we discuss collaboration with external partners.

19. See Ward et al. 1995.

20. Studies of innovativeness have repeatedly shown that group rewards emphasize execution, while individual rewards emphasize idea generation and radically new ideas. See, for example, Angle 1989, p. 142. In addition, incentive experts in economics have shown that when employees are confronted with multiple, partially conflicting tasks, strong incentives push them toward the less uncertain and more predictable tasks, because in this way, they can better guarantee some output and a positive evaluation for themselves. In economics, this is referred to as the impossibility of strong incentives for multiple tasks with uncertainty (Holmström and Milgrom 1991). In projects with unk unks, it is, of course, exactly the uncertain tasks that must be tackled.

21. This was first proposed by Sinclair-Desgagné 1999. See also Sommer and Loch 2005b for a discussion in the context of unk unks.

22. See, for example, Pfeffer and Sutton 2000.

23. We know this from empirical studies in innovation research; for example, see Angle 1989, p. 142. In addition, there is much research on incentive systems that shows this see, for example, Kohn 1993, Kunkel 1997, Pfeffer 1998.

24. See an overview of the reasons and the implications in Loch, Yaziji, and Langen 2000.

25. This discussion is based on Terwiesch, Loch, and De Meyer 2002; the rest of this paragraph is quoted from this article, p. 402.

26. Kitchen building is, by the way, a good example of a project in which unk unks can easily occur. As the aesthetics of the finished kitchen depend on subtle and complex interactions of colors and other elements, the final look is very difficult to predict and may easily come out differently than was intended. The design space is "unstructured" and must be searched (see Terwiesch and Loch 2004).

27. See Terwiesch, Loch, and De Meyer 2002, p. 409.

28. Adapted from Terwiesch, Loch, and De Meyer 2002, p. 412.

29. For a discussion of principles according to which a course of action can be chosen in dealing with the preliminary information, see Loch and Terwiesch 2005.

Managing
Unk Unks
with Partners

Major projects are often, and increasingly, carried out by combinations of firms and public institutions, as they involve risks that few
organizations are able to take alone.[1] This is all the truer if a project has
novel aspects and involves unknown unknowns.[2] A partnership with an
external party introduces relationship complexity (see Chapters 3 and
4): Control over external parties is often not direct, but only possible
through contracts or persuasion; project participants from different
organizations possibly have objectives and priorities that are different
from, and sometimes even conflicting with, those of the project owner;
and they may use different jargon and have different ways of dealing
with collaborators (different cultures). In a coalition, a frequent consequence of unexpected events hitting a project is its unraveling; the project disintegrates.

The higher the number and interdependence of the parties involved, the more difficult is the management of the relationships. Obviously, in an age of increasing outsourcing and dispersed expertise, this is often correlated with the complexity of the project tasks; the more dispersed the task and the expertise, the more parties will be involved.

The project management infrastructure that we described in Chapter 9 is not sufficient to handle external partners. This is because the management infrastructure assumes that the relationship between project management and the project team is "open-ended": Essentially, the team does everything that management demands (within reasonable limits, of course), for as long as is necessary, in order to bring the project to a successful conclusion.

However, the principal instrument that is used to bring about cooperation and coordination with external parties is a form of contract. We discussed contracts in Section 3.3.2. They are not of the above-described open-ended nature;[3] rather, they outline what each party is supposed to do and what it is not supposed to do. Contracts cannot specify what the parties should do after an unk unk occurs, because unk unks cannot be foreseen. Contracts *can*, however, describe the *process* that is followed to handle an emerging unk unk. If the interaction among project participants is governed by a contract, other ways of adjusting to unk unks must be found.[4]

We demonstrate the limitations of contracts in Section 10.1. Then we describe five principles of partner management in order to ensure a constructive handling of unk unks in the project. We summarize the implications of these principles at the end of the chapter.

10.1 The Dangers of Project Contracts
10.1.1 The Eurotunnel Project

The Eurotunnel, running under the English Channel to connect the British Isles with the continent of Europe, near Calais, is famous for its budget overruns and subsequent shareholder tensions.[5] The idea for this tunnel was resurrected in 1984, after at least 26 previous schemes, the first in 1802 and the last in 1978, had fallen through.[6]

The tunnel traverses 26 miles under the channel between Folkstone on the British side and Calais on the continent of Europe. It comprises two parallel tunnels with a service tunnel in the center and two crossovers between them (Figure 10.1). The depth profile of the tunnel is shown in Figure 10.2.

The Eurotunnel example is not so much about major fundamental unk unks. Both the tunnel technologies and the passenger and freight transportation markets across the channel were basically known (as we further explain below). Rather, it is an example of the relationship between complexity and uncertainty. The system was so complex that significant residual

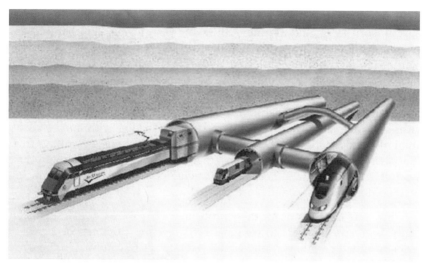

Figure 10.1 Tunnel architecture (Copyright: Eurotunnel; reproduced with permission.)

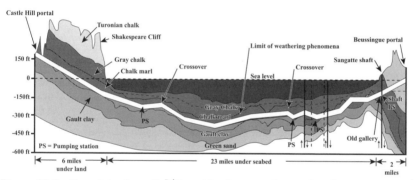

Figure 10.2 Tunnel depth profile[7] (Copyright: Eurotunnel; reproduced with permission.)

uncertainty[8] could not be excluded. In addition, the degree of relational complexity was so high, and the relationships so dysfunctional, that the partners were not able to deal with changes caused by residual uncertainty. Worse, the complex relationships even *caused* unk unks, in the sense that, later in the project, players were completely taken by surprise by the consequences of actions of other players. Contracts were completely inadequate to deal with the combined effects of unk unks and conflicts of interest.

The European Commission was keen on improving the European transport networks, the railway utilities wanted to increase their attractiveness with a connection from Paris to London, and the private sector was hungry for a major project. A group of contractors and banks set up the Channel Tunnel Group, together with a mirror contractor group on the French side. In 1985, the consortium submitted a bid in response to an invitation document issued by the French and British governments.

In short, digging started in 1987. The tunnel opened for freight transport in April 1994, and in June of the same year, for passenger trains, one year late. This schedule overrun is not the key problem (especially in the light of the fact that this reflected a six-month schedule tightening, midway, under external pressures). The key problem was a budget overrun from $7 billion to $13 billion. As a result, the outcome of the project has not been economically viable: Eurotunnel, the operating company that owns the operating license for 55 years, was loaded with such massive amounts of debt, plus some operating cost disadvantages resulting from shortsighted tunnel design decisions, that it has never been able to produce a profit and has undergone two refinancing crises, in which the current shareholders have essentially lost money.

What were the reasons for these problems? They were not market uncertainty: When we compare the original 1987 revenue estimates for 2003 (10 years into the operation) to the actual revenues of Eurotunnel in the company's 2003 annual report, they are within 10 percent of each other.[9] The market analyses correctly estimated the market potential, including the competitive response by the channel ferries, which drove prices down by 50 percent when the tunnel opened.

The technological challenge did indeed "push the state of the art" in size, although not fundamentally so. For example, the huge tunnel-boring machines were based on previous designs but had to be made more sophisticated to allow for varying boring widths. These were required because the concrete lining segments of the tunnel (each segment was a very wide, 9-meter-long pipe) had to be of varying thicknesses, depending on the surrounding ground pressure, and therefore, the excavation volume also had to vary. In addition, more water than expected was found under the chalk land of the channel, which necessitated modifications to the boring machines. This caused no more than a one-month delay. Otherwise, however, the tunnel design was based on existing technologies, only larger. In particular, the tracks, signaling, and trains were intended to incorporate standard TGV train technology and should not have posed any major unplanned challenges.

Overall, the technological challenge does not seem to justify the actual delays and overruns. Indeed, the project used sophisticated but standard scheduling and planning software, for which it was commended (see Figure 10.3).[10] The problems did not stem from inappropriate planning.

Subsequent studies have revealed that the problems overwhelmingly stemmed from the relational complexity of the web of actors, and their conflicting interests, being influenced by the contractual arrangements.[11] These contractual arrangements were never designed in the best interest of the overall project but were substantially fixed at the outset, in a context of political lobbying, even before Eurotunnel, the operator and owner, was founded. They then evolved over time as a result of ensuing power struggles. Moreover, there was no master project manager and no one who oversaw the entire complex web. Therefore, actions by some players had completely unforeseeable consequences for other players; in other words, the unk unks resulted from the relationship complexity more than from market or technological novelty. Figure 10.4 summarizes the actors and their interests.

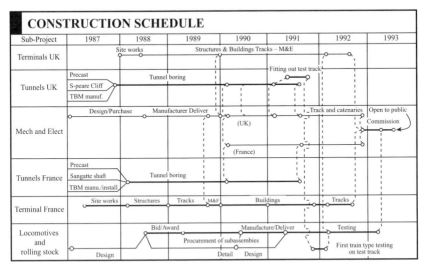

Figure 10.3 Overall project schedule and planning

(Source: VF, *Civil Engineering* 1989; reproduced with permission by ASCE.)

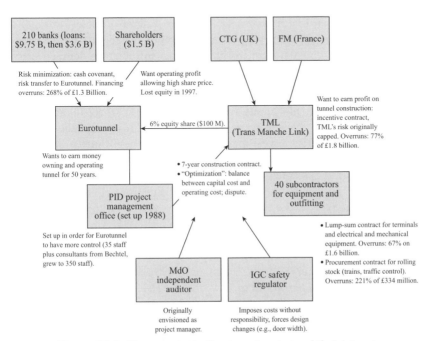

Figure 10.4 The actors in the Eurotunnel project and their interests

The two governments wanted the tunnel to happen but insisted (in the spirit of the 1980s) on private financing and ownership; in particular, they refused to give any guarantees to the banks, which inevitably increased their risk and reduced their enthusiasm. The governments promised to build a high-speed rail link in their respective countries to the edges of the tunnel but did not fully deliver. The French government finally did so, in time for the opening of the tunnel, but the British government failed to deliver at all; in fact, half of the link between London and Folkstone was opened only in 2003, and at the writing of this book, the other half is still missing (this still increases the travel time from Paris to London from two to two and a half hours today).

The first negotiations, and the design of the tunnel, were performed by TML ("Trans Manche Link"), the Franco-British consortium of construction companies. TML[12] won the bid to construct the tunnel. Eurotunnel was not formed until 1986 and was never fully accepted by TML, although Eurotunnel was the owner of the operating license and was formally the project owner and TML's client. TML saw Eurotunnel as a "Johnny-come-lately, whose responsibility was to pay TML's monthly bills and nod to the design."[13] TML wanted the design not to run over budget but was not concerned about Eurotunnel's later operating costs; a number of design decisions were made that increased Eurotunnel's maintenance costs (for example, the costs of three pumping stations that pump water from sumps to treatment plants at either end; ventilation and cooling systems; and the final train speed, which is further discussed below).

The original contract (negotiated before Eurotunnel was formed) foresaw that TML would be responsible for only 30 percent of overruns on the tunneling works, and with only a small upper limit. It took Eurotunnel a year to set up a competent project management office that could effectively oversee the construction works and start influencing detailed decision making as it happened. Relationships between Eurotunnel and TML remained adversarial throughout.

When Eurotunnel placed its first equity offering of $200 million in 1986,[14] the financial markets were reluctant to pick it up, and the Bank of England and the British government lobbied (interview sources later used the term "bullied") to persuade banks to take up the equity. As a result, the banks became extremely cautious and risk-averse. First, they placed a covenant in all loan agreements that required Eurotunnel to have enough cash on hand to pay for the entire project all the way to completion (to prevent any possibility of a default at a point when there was a half-complete hole in the ground that was good for nothing!). This caused a major crisis in 1990, when Eurotunnel's reserves fell short. This almost led to the project's demise and caused construction work to stop for a month. Eventually, the banks lent more money, and the risk balance between TML and Eurotunnel was changed to an equal sharing of overruns without an upper limit. However, risks were still predominantly on Eurotunnel's side. Interest and inflation risks were borne by Eurotunnel alone, and financing costs experienced the biggest percent overrun of all cost categories (268 percent, i.e., from £1.3 billion to £3.5 billion, see Figure 10.4).

Subcontractor relationships also caused problems. TML's competence was in tunnel construction; they did not really want to carry responsibility for track construction and railway systems engineering. TML were forced into a general contractor role because the banks wanted to have one responsible counterpart to talk to (this was settled before Eurotunnel's formation). Consequently, TML minimized management attention to those unwanted parts of the project, a move that caused difficulties in two major areas.

First, the construction of the terminals and installation of electrical and mechanical equipment was thought to be straightforward because only proven technologies were to be used. Thus, TML accepted a lump-sum contract and, in turn, imposed lump-sum agreements on the subcontractors. This was "frankly naïve," as one observer called it, and led to claims by TML for the cost of changing designs and specifications, partly at the behest of the Inter Government Commission (IGC), but also because the complexity of the systems had been underestimated: There were interactions between the design of the fixed equipment and modifications in the tunnel and the trains. A court battle between TML and Eurotunnel ensued, which exacerbated the crisis in 1990.

The procurement contracts for the rolling stock (costs were rolled through, and TML received a fixed management fee) constituted an abdication by TML, who felt that they had insufficient expertise and were pressured into the overall responsibility by the banks. Reflecting the lack of expertise, subcontracts were signed such that there was little competition for the rolling stock work, which led to large overruns there as well.

Conflicts between Eurotunnel and TML simmered throughout, over "optimization," a provision of the original contract that stipulated the achievement of the "best balance between capital cost [which TML wanted to minimize] and operating costs [which Eurotunnel wanted to control]." Participants stated in interviews that "Eurotunnel and TML were in total disagreement on how to interpret the concept of optimization." A number of design changes worked in TML's favor; for example, when rolling stock costs spiraled out of control, the maximum speed of the trains was limited to 80 km/h in order to save costs. This doubled the tunnel passing time, reducing Eurotunnel's capacity and competitive advantage and adding to its operating costs.

The final player to be mentioned was the IGC, the government supervisory body, which had been set up to coordinate the British and French governments' policies concerning the tunnel's construction, operation, and safety. It imposed major safety-related changes to the fixed equipment (power supplies, tracks, mechanical systems, etc.) designs submitted in the 1987 agreement. It also caused major cost increases in the rolling stock. For example, it delayed the approval for a safety-related proposal of widening the passenger car doors from 60 to 70 cm. When the approval did not come, TML continued manufacturing to keep on schedule. But later, the IGC decided that 60-cm doors were unacceptable, and the option had to be installed anyway, but at this point, the decision had caused a nine-month delay and a £45 million cost increase.

The IGC clearly prioritized safety over cost and speed. Eurotunnel and TML (this time in concert) complained that the IGC had "the authority to change and control things without commensurate responsibility," taking away necessary margins and reserves and causing delays that made costs skyrocket.

Summing up the Eurotunnel example, relational complexity of this project was high, and the contractual structure, which emerged through political haggling over time, pitted the parties against one another, so constructive problem solving during the project became very difficult. Worst of all, the complexity of both the technical systems and the relationships even *caused* unk unks, in the sense that players were completely taken by surprise by the consequences of actions of other players (e.g., the IGC's design changes; TML's claims regarding fixed equipment and rolling stock, which had been judged straightforward; etc.).

In addition to the interest conflicts, the parties were not able to build constructive relationships; there were clashes among the senior managers on the different sides, and the organizations, over time, assumed a "winners and losers" mentality.[15] This exacerbated the dysfunctional contract structure, precisely when problem solving and changes were required, and the parties dug in their heels to prevent adjustments. This example illustrates how contract structures that work in simpler projects can lead to failure in complex projects with considerable uncertainty.

10.1.2 Other Examples

In order to appreciate that the Eurotunnel example is not a grossly deviant exception, consider a second example, the South Trunk project,[16] an independent power plant project designed to burn waste coal from a nearby pile. The main risk that managers anticipated was technical difficulties with the boiler that used the relatively new and untested circulating fluid bed technology.[17] Their worries proved to be unfounded, and the boiler, for which a reputable supplier was selected, did not cause any significant problems during startup and operation.

However, the South Trunk project *did* experience significant problems related to the fuel-handling system and to a reversal in the trends for fuel and electricity prices, none of which had been anticipated by the participants. During startup, the project experienced repeated failures of the waste coal- and ash-handling systems. Participants started blaming one another. The owner blamed the turn-key contractor, who, in turn, blamed the owner because the coal received from the nearby waste pile contained higher humidity than specified in the turn-key contract. An unexpected decline in coal prices ultimately led to the demise of the project.

The payments that the project received from its utility client were tied to the cost of the coal-fired power generation in the utility's plants. Technical difficulties became pretexts for each party to jump upon for an opportunity to exit with minimal losses. After making costly modifications to the project and persuading the owner of the waste pile to change the contract and bring in higher-quality coal from the outside, the bank took over the project but failed to operate it efficiently. The bank finally sold the project to the client utility at a quarter of its cost. The utility shut it down, arguing that it no longer needed the capacity.

In summary, the participants were no longer able to overcome (seemingly) differing interests, and cohesion, or the capacity of the participants to collaboratively solve problems caused by unforeseen events, broke down. As a result, the project failed.

Examples of contracting problems in novel projects abound, and not only in the construction industry. To give an example in the electronics industry, the automated interstate highway truck toll collection system introduced in Germany in January 2005 was delayed by 18 months, an overrun of 50 percent. The system was highly complex, combining on-board GPS/infrared/radio units with in-built software (storing a digital map of the country's road network and interacting with the central system), a coordination satellite, road toll sensors, and central data processing, all of which would allow not only toll collection but also intelligent vehicle routing, according to traffic conditions. This complex combination of newly developed components became a massive problem because the contractual structure weakened the parties' motivation to coordinate and collaborate: The contractor itself (TollCollect) was a consortium of three firms, which made decisions by consensus. This setup led to an inevitably slow process, and when problems arose, the partners started to shift responsibilities and blame one another. Second, the contracts between the consortium and the government produced unclear incentives. Third, the politically motivated inclusion of a third party, one of the losing bidders for the system, in manufacturing the on-board boxes, allowed the consortium to escape contractual penalties. Again, the contractual structure significantly exacerbated the challenge of dealing with the unk unks that, in this case, arose from the complex combination of novel components.[18]

10.1.3 The Limitations of and the Need to Extend Contracts

The above-described problems are not exceptions; rather, they are commonplace.[19] The Circored project (Chapter 2) also suffered from contract confrontations, when the general contractor, Bechtel, rejected bids for the EPC construction contract because Bechtel thought it could be done at lower costs. Bechtel then performed the EPC contract itself, in a fixed-price structure, but found out that it had underestimated the costs. Subsequently, a legal confrontation ensued, during which Bechtel filed many claims demanding more money.

It is a quite widely used strategy of contractors to bid low and then file legal claims for minor changes to the activities (narrowly interpreting what the contract demands). This has been observed in multiple industry contexts of project management, leading some authors to conclude the following:[20]

> A contract is a dangerous instrument and should always be approached with trepidation and caution. . . . Theoretically, the aim of a written contract is to achieve certainty of obligation of each party, the avoidance of ambiguities, and such definiteness of understanding as to preclude ultimate controversy. In practice, construction contracts are generally formed not to definitely fix obligations, but to avoid obligations.

More generally, one hears the observation, "The more predictable construction environment of the past has given way to a massive number of unknown, unpredictable, and unquantifiable problems. This change has resulted, in part, from the growing number of relationships in modern projects. These new relationships tend to produce conflict, not cooperation."[21]

We therefore arrive at the inevitable conclusion that traditional contracts are insufficient to ensure collaboration of multiple parties in a project. Aligning behavior is difficult enough in projects with high relational complexity without unk unks. But as the examples and assessments by professionals show, unk unks make it hopeless to specify contracts on concrete activities and deliverables. When unexpected changes occur, partners are affected differently, invalidating carefully tuned contractual agreements, and the project inevitably falls apart.

The more carefully a contract attempts to foresee contingencies and regulate risks, awards, and behavior, the more dysfunctional it becomes when unexpected events change the project. The very measures taken in order to stabilize the future and avoid anticipated risks reduce the flexibility of a process that governs the problem solving in response to unk unks[22] because the players try to hold on to what they have; detailed contracts irresistibly prompt players to block change.

10.2 A Problem-Solving Process in the Face of Unk Unks

In novel projects subject to unknown unknowns, contracts must be embedded in a system of alignment and coordination measures. Only then has a novel project a chance of overcoming the inevitable tensions associated with major changes. Maintaining the ability for joint problem solving in the face of changes requires establishing a mutually agreed upon and shared *process* of problem solving.[23] The steps of partner management are as follows: Choose partners to assemble a collection of competences, clearly allocate risks and rewards and maintain flexibility in the details, apply fair process in problem solving during the project, install a transparent early warning system, and build relationships with the partners over time, enabling collaborative problem solving. We describe these five steps of partner management in this section.

10.2.1 Choose Partners for the Competences They Contribute

It is commonplace for project contractors and partners to be chosen based on price. In other words, the lowest bid gets the contract. This widely used practice has led to the equally widely used counter-strategy. "Usually, project management holds a beauty parade, and takes on the suppliers who bid lowest. The suppliers rely on glitches and delays to bump up the costs. Every time something goes wrong, legal haggling breaks out among suppliers and between them and the owner, work shuts down for weeks on end, and a huge slice of the costs ends up in the pockets of lawyers. Once the construction is late, time runs short for the final installation and testing of the electronic systems."[24] Such mutual gaming can work in projects with little uncertainty, but it will spiral out of control in novel projects where changes are inevitable. In novel projects with unforeseeable uncertainty, two criteria should be judged as important as price, or even overriding it, in the choice of project partners: competences and relational compatibility.

Competence Bundles

Throughout this book, we have discussed the fact that, in the presence of unk unks, the project plan cannot specify tasks; tasks are unknown and the plan is only a stake in the ground. Thus, contractor choice based on a bid purely on specific tasks is an illusion. Instead, once the project areas in which unk unks loom have been identified, project management should ask: *What are the competences that we need in order to "cover" the areas of concern?* What competences do we need to be able to effectively respond to unexpected events in the areas of concern, whatever they are?

After this clarification, the partner who has the *deepest competence in this area and a track record* of problem solving and performance should be the one chosen. Once project management knows that unk unks threaten, any initial price advantage is an illusion and is likely to be dominated by the costs of adjusting the project later. Effectiveness and costs of adjustments are driven by bundles of competence. The choice of a group of partners determines what competences the project management team later has at its disposal.

Of course, project management professionals have known this for a long time. And yet it is often not done because the temptation proves irresistible for upper management to choose cheaper bids, and go for immediately visible "savings," often by overriding project management.

Relational Compatibility

In addition to competences, the ability to work together has a big impact on project success (we will discuss this further in Section 10.2.4). Project management should meet important contractors face-to-face to understand the chemistry and mutual attitudes.[25] Four specific "initial conditions" that influence the chances of the partners achieving a constructive working relationship should be checked: the common understanding of the task definition, the partner's organizational routines, the interface structure, and mutual expectations of performance, behavior, and motives.[26] We explain each in turn, illustrating them using the example of the Circored project from Chapter 2, specifically, for the initial conditions for a collaboration between Lurgi (the technology owner of the core plant), and Bechtel, the construction company of the facility (before startup).

1. *Task definition.* This is what the shared project is designed to achieve and what each side brings. Lurgi had bid to become the general contractor, but Cliffs chose Bechtel. Bechtel saw Lurgi strictly as a subcontractor for the core plant and attempted to minimize Lurgi's scope of activities as much as they could. Lurgi saw the entire first-of-a-kind facility as their baby. This resulted in continued disagreements when who strayed on whose turf.

2. *Organizational routines.* Different organizations have procedures, cultures, and "ways of doing things" that are executed automatically and not always consciously recognized. Significant differences across partners cause friction and make collaboration action more difficult. In the Circored project, Lurgi was a small engineering company that survived by offering clients facilities that were, at least, partly customized. In contrast, Bechtel was a large construction company that competed on standardization and cost reductions. Thus, Bechtel was accustomed to working by highly prescriptive processes, even the slightest deviation from which had to be authorized via a change request. Lurgi operated under the assumption that no two solutions were ever the same; thus, their personnel had the authority to tinker, even the front-line workers (all of them skilled). Lurgi accused Bechtel of inflexibility, incompetence, and causing unnecessary delays in granting work authorization. Bechtel viewed Lurgi personnel as unprofessional, never doing anything by the book, or in any consistent or reliable way. Interminable clashes and tensions during the project were the result of these incompatibilities.

3. *Interface structure.* The interface refers to how many people are involved on each side and how often they interact. If the interface is shallow and infrequent, any joint knowledge generation will be slowed down because the partners do not meet often enough to develop customized ways of collaborating. If the interface is disrupted, any tacit knowledge that has been built up, which cannot be written down in memos and minutes, is lost. In the Circored project, the Bechtel project manager was replaced three times over the two years, and Cliffs' overall project manager was on-site only one week per month. This made close collaboration difficult, to say the least.

   Often, the interface must work at different levels of the organization simultaneously: Senior managers must agree on the strategic aspects of the project, while technical experts have to collaborate to solve operational problems. Thus, interfaces must exist at multiple levels. This can become difficult when the two organizations are of very different sizes or organizational designs. On the other hand, multiple interfaces can also make the collaboration easier, as tensions at the top (haggling over cost or benefit sharing) may coexist with close cooperation and problem solving at the operating level.

4. *Mutual expectations and compatible goals.* Each partner enters the collaboration with explicit and implicit expectations, and with hypotheses about the other side's motives. The goals and expectations of the two sides do not have to be common and shared, but they must be compatible—one side getting what it wants should not be incompatible with the other side also having its goals fulfilled. In other words, if the partnership is a zero-sum game, collaboration becomes much more difficult.

Expectations sometimes become self-fulfilling prophecies. In the Circored project, Bechtel thought Lurgi was arrogant, as if Lurgi thought they were the only ones who understood circulating fluid beds. Lurgi, in turn, warned Cliffs that Bechtel always underbid and then went after change claims. The two companies had worked together on a previous project, with negative results. Although the parties involved had been a different Bechtel office and a different Lurgi department, the past experience caused negative expectations on both sides.

In the Circored project, the initial conditions were so negatively loaded for the collaborating parties, Lurgi and Bechtel, that they could not be overcome. Collaboration between these two contractors never recovered. To understand that goals do not have to be common, that compatibility is enough for a fruitful collaboration, consider the project of constructing INSEAD's Asian campus in Singapore between 1998 and 2000. An important partner in this project was Singapore's Economic Development Board (EDB), an organization that established connections between INSEAD and local constituencies and supported real-estate transactions and helped establish a research fund. The two organizations had different goals—the EDB wanted to contribute to a thriving and competitive academic environment in Singapore, while INSEAD wanted a commercially and intellectually fruitful beachhead for its organization. The basis for a successful collaboration lay in the fact that one party's set of goals could further the other's set of goals.[27]

Through background checks and personal contacts, it is possible for project management to sound out these initial conditions for important contractors at the beginning of the project and use them as a choice criterion in addition to the competence bundle desired.

10.2.2 Clear Risk and Reward Allocation and Flexibility

We started Chapter 10 by criticizing contracts and arguing that they do not suffice. However, we are, of course, not recommending eliminating contracts. They must be embedded in additional ways of encouraging collaborative behavior, while playing a central role in shaping the collaboration. The contract defines the business deal and sets the tone of the relationship. In this section, we argue that contracts must clearly allocate risks and rewards, be flexible, and be complemented by interest alignment via co-ownership if unknown unknowns are major.

Allocation of Risks and Rewards
Professionals commonly call for contracts to have three characteristics: "All risks should be considered as belonging to the [project] owner unless specifically assigned to another party by the provisions of the contract. . . . Determining who should be assigned a risk should be based on who has the competence and expertise to deal with that risk."[28]

The contract should not be a *political* document that clouds difficult issues in diplomatic language, but a document that supports PRM by clearly spelling out the risks.[29] As we discussed in Chapter 3, the contract defines the business deal, and the clearer the risks and responsibilities are spelled out, the more constructively the parties can behave later.

Indeed, one empirical study suggests that contract usage in many projects reflects the parties' ability to deal with the risks.[30] This study examined whether projects used outcome-based contracts (fixed-price, in which all performance deviations are borne by the contractor) versus "behavior-based" contracts, in which the price depends on other considerations (presumably, effort and process quality by the contractor). The study showed that behavior-based contracts were used more when the contractor was small (and thus not able to absorb large risks), when interests between client and contractor were highly aligned, when the client was highly competent (and thus able to closely monitor the contractor's behavior), and when the project was novel (and thus unforeseeable changes were to be expected, which were outside the control of the contractor). Outcome-based contracts, in contrast, were mostly used when the client was small and unable to absorb large risks.

Contract Flexibility
It is well known that a high likelihood of changing requirements, cost uncertainty, and difficulty to measure performance prevents parties from drawing up "complete" contracts that include all contingencies. In other words, complexity and uncertainty thwart complete contracts.

Therefore, contracts in novel projects need elements of "hierarchy," or de facto oversight and decision structures as if they were within the same organization.[31] In other words, detailed actions to be taken and detailed outcomes are *not* specified in the contract (although the general areas of responsibility and the general nature of the desired outcome are); rather, the contract defines an "open-ended" characterization of work, analogous to an employment relationship, in which the contractor is expected to execute orders and contribute to the project activities in a way that the owner deems satisfactory. An open-ended agreement gives the flexibility to respond to unknown unknowns.[32] Specifically, this open-endedness applies to changes in the specifications, which are inevitable in a subproject with a potential for unk unks, provided that the contractor is reimbursed for costs of the changes.[33]

Finally, it is well known that the contract should always contain dispute resolution mechanisms as "ways in which the parties air and resolve differences about the interpretation and performance of the contract; . . . they may be thought of as 'grievance procedures.' . . . The purpose of dispute procedures internal to the contract is to prevent minor disputes from developing into expensive and disruptive legal battles."[34]

It must be emphasized that dispute resolution procedures in a project with unk unks must not only be intensified but must take on a whole different character. They must be elevated from the above-expressed spirit of a "depressurizing valve of last resort" to normal, everyday procedures of shared problem solving. If our plan is only a stake in the ground, and we have to evolve it as we proceed, we must be able to collaboratively make changes and solve new problems every day, without the threat of a dispute or legal action in the back of our minds. We must be able to resolve inevitable differences in our views and the interpretations of observations routinely and collaboratively. Thus, the term "dispute resolution mechanisms" should be replaced by "shared problem-solving mechanisms."

Co-Ownership

It is unrealistic to hope that classic client-contractor contracts can completely solve the challenge of unforeseen contingencies, such as a complete failure of the technology. These contracts simply fall outside the traditional supply contract toolbox. Under high uncertainty, *additional* interest alignment is necessary by sharing ownership of the project, for example, by running it in a joint venture co-owned by the contract partners. This is, of course, easier if the parties' goals are compatible from the outset.

Returning to the Circored example, the collaboration between Cliffs and Lurgi was facilitated by the fact that Lurgi owned 7 percent (and later 18 percent) of the joint venture, CAL. This, at least, gave both sides minimum incentives to keep working during the first major crisis in the summer of 2000, when the external consultant recommended shutting the facility down.

Even here, however, caution is warranted; co-ownership may leave the parties with different levels of exposure and priorities relative to the size of their respective businesses. The party for whom the project is less important may still decide to abandon the party that depends on it more. Coming back to the Eurotunnel example, TML's 6 percent ownership of Eurotunnel was simply insufficient as a motivation for TML to help Eurotunnel with its debt load and operating cost structure.

10.2.3 Fair Process

John, a project professional whom we know, was given a bad performance evaluation by his manager, which he angrily disputed. He had a meeting with the manager, in which, after a shouting match, the manager agreed to change his mind and to give John a higher performance rating. After he emerged from this meeting, John fumed, "It is amazing how this guy has the ability to, in the end, give you what you wanted, but in the process still make you pissed off." And from then on, John proceeded to undercut the manager, in subtle ways, wherever he could.

The outcome of the evaluation was what John had wanted. But he hated the process of getting to the outcome: He found it unfair. The example illustrates a general desire that people have: They want fairness. They like positive outcomes, but the positive outcome is galled if the process was perceived as unfair, and even a negative outcome can become palatable if the process was fair and just.

The desire for fairness is a universal and deep psychological human need. We are social animals, and we care about justice in our social group. If fairness has been violated, we feel anger and indignation if it has been done to us, and shame and guilt if we have done it to others. More than that, we even have an in-built "cheating module" in our brain: In situations where social cheating (violation of fairness by taking advantage of the other party) is possible, we automatically become highly alert and, unconsciously, scrutinize information very carefully to see whether an unfair act has indeed occurred.[35]

Often, we hear managers complain that their employees and business partners always second-guess them, even when they think they have communicated clearly. But gossiping and second-guessing are not signs of irresponsibility of the masses. On the contrary, they reflect a healthy, cautious attitude to situations of vulnerability, an "instinct" of wanting to scrutinize people with power whenever there is the slightest possibility of abusing that power. Thus, the following three behaviors are normal, rather than deviant: a concern for fairness, relentless suspicion about fairness being abused, and a readiness to act violently when an abuse of fairness has indeed occurred (most of the time, only in a figurative sense, but sometimes literally).

Fairness is very important for novel projects. If unk unks emerge and the management team is forced to modify the project plan, opportunities for taking advantage of partners are rampant. It is very difficult for the partners, who do not have complete information, to judge whether the modification had to be so drastic, whether so much of the change had to be shouldered by them, or whether the change had to come just now. If fairness is not convincingly demonstrated, the natural and understandable reaction is suspicion, protest, and possibly blockage, or at least a subtle withholding of the best effort. How can the project team prevent this reaction?

The Structure and Effect of Fair Process

Fair process has three principal parts.[36]

1. *Clarity of expectations.* This means the clarity of the rules of the game, of the overall purpose, and of the performance that must be achieved. When people clearly understand what must be achieved and where their contribution lies, political jockeying is reduced, and the participants can focus on the job at hand. This includes the credibility of the project owner in setting the expectations.

2. *Engagement.* This means involving the affected individuals in the decisions that concern them by asking them for their input and allowing them to refute the merits of one another's ideas and assumptions. It communicates management's respect for these individuals and their ideas. Engagement is embedded in a context of regular mutual communication, which allows the parties to know one another and to understand how they think and argue, and which prevents the discussion of the decision from coming out of the blue.

3. *Explanation.* The reasons for the change and why the project had to be modified this way must be laid out clearly. The explanation must make the reasons transparent and demonstrate that no hidden agenda or "secret deals" are involved. The clarity and transparency works against the automatic suspicion (the "cheating module") and engenders trust, even if the idea of a partner has been rejected. Engagement and explanation also serve as feedback loops that enhance learning, both on the project management's side and on the side of the partner or contractor.

The effect of fairness on behavior can be considerable. Studies have been performed on fair process not in project management, but in the context of organizational performance. One study analyzed the behavior of international managers in response to centralized decisions made by the head office. The study found that when fair process was followed, decisions were more easily accepted, the subsidiary managers collaborated more, and moreover, they volunteered their own initiatives and ideas that enhanced the decisions and improved the organization's learning.[37]

Why does fair process make such a big difference in people's behavior? The reason is not rational calculation of benefits, but it is an *emotional* affair. Paying people respect by asking their opinion, and deactivating the automatic subconscious "cheating module" by providing them with transparency allows them to trust instead of second-guessing, and it makes it possible to accept even uncomfortable changes with their head held high. This gives a small emotional push to *wanting to collaborate*, as opposed to *wanting to get even* if fairness is violated. Fair process does not override incentives. If I lose from the project change while others gain, I will be against it and fight it, fair process or not. However, there is a large gray area of outcomes where fair process makes a great difference.

For example, in one offshore oil platform project in the mid-1990s, management invested heavily in collaborative partner relationships. But in early 1994, oil prices dropped by one-third, which made the entire project unprofitable. In order to rescue viability, everyone needed to make some concessions. And everyone *did* contribute. One contractor delayed the start of jacket fabrication by seven months, deferring expenditures of £10 million. The general contractor committed to staff reduction without compromises in performance, saving 10 percent on overhead costs. A construction subcontractor offered design cost reductions by reusing lifting beams designed for an earlier platform. And so on. This set a tone of collaboration and compromises, initiated a stream of changes and adjustments, and kept the project alive.[38]

While fair process sounds great and *does* make a difference when followed, it is difficult to put into practice, for two main reasons. First, fair process makes pursuing hidden agendas much harder. If an opportunity arises, the temptation of taking advantage of it at the cost of the other side may prove irresistible. After all, many situations in a project represent a zero-sum game in their immediate effects: Either I win and you lose, or vice versa (although in the longer run, win-wins are more sustainable). Therefore, the reluctance to open up is great, and a complete "opening up" is rare. For example, the Heathrow Airport T5 (terminal 5) project, which is ongoing as we write this book, has publicly stated that it wants to collaborate with its suppliers. The project owner, the airport company BAA, has agreed to carry all risks, putting a large contingency reserve of funds aside that will be shared among the suppliers. However, even here, this agreement applies only to a small subset of suppliers and has not yet been put to a real test.[39]

Second, engagement and transparency open up the possibility of being wrong, and that is threatening. If I allow engagement, I open myself up to the other side finding an error in my logic, and then I will have to agree to some modification, which may make elusive the solution that I would really like. Fair process requires honesty and the self-confidence to be able to admit to being in error, and then to look for an alternative with the other side. Frankly, many managers simply do not have this level of security.

10.2.4 Early-Warning Systems

We want to elaborate a bit more on one dimension of fair process: transparency. An important aspect of transparency is an early-warning system, or the systematic communication to the other party of the degree of uncertainty of information, and of early signs of unexpected events or problems. If unexpected changes emanating from a partner are indicated as they are emerging, and if their reasons are understood, trust building and stability of the relationship become greatly enhanced.

Early-warning systems place requirements on both partners: the willingness to release the information, and the willingness to receive the information and to respond to it. We discussed systems of exchanging preliminary information among subprojects in Section 9.4.3. In effect, such a system needs to include not only internally staffed subproject teams but also external partners. The effect of open preliminary information transfer lies not only in the effectiveness of PRM, as discussed in Chapter 9, but also the robustness of the partner relationship in the face of unexpected shifts in mutual interests.

10.2.5 Relationship and Trust Building

Even with a good contractor choice, a flexible contract, and decision making that follows fair process, changes in the plan being forced by emerging unk unks may be so painful that some parties give up. So, the project may still disintegrate. To maximize the chances of maintaining constructive problem solving, the relationship between the parties must be developed and invested in throughout. And this requires mutual adjustment, not only adjustment by one side (no matter how well backed up by fair process).

The power of mutual adjustment has been documented in a study of strategic alliances.[40] Strategic alliances are similar to novel projects, as they are well defined in their scope of collaboration, and they often have a defined end—the parties engage in the alliance in order to gain access to a certain market, or to acquire certain knowledge. Often, alliances are terminated after a few years when those objectives have been achieved (or when the parties realize that the objectives are not achievable). Moreover, alliances are almost always affected by unk unks, for several reasons: Usually, they are formed to tackle new markets (that neither partner can address alone), or one partner wants to learn something new. Moreover, the organizations discover each other and learn how to deal with each other. As a result, alliances always feature learning and modifications in response to unk unks.

Figure 10.5 summarizes typical differences between successful alliance projects and unsuccessful ones, by focusing on key dynamics of the *process* of the collaboration. The project starts with the initial conditions that we have discussed in the preceding sections. The initial conditions load the dice for the chances of success of the project. Then the parties enter the learning path of the project as unexpected findings emerge.

Learning comprises two aspects. The first aspect involves the learning that we have discussed throughout this book (referred to as "content learning"), knowledge about and responses to the environment, and the project's success drivers. Content learning also includes learning about (and interpreting) the partner's hidden motives in the project. Second, the ability to adjust one's behavior in the interaction with the partner matters. For example, does the partner manage to adjust its organizational routines to facilitate interaction? Is the partner willing to change reporting routines, or travel authorizations, or decision-making rules? Does the partner engage in reevaluations of the business plan? Are additional resources and people brought in if it helps the project? And so on.

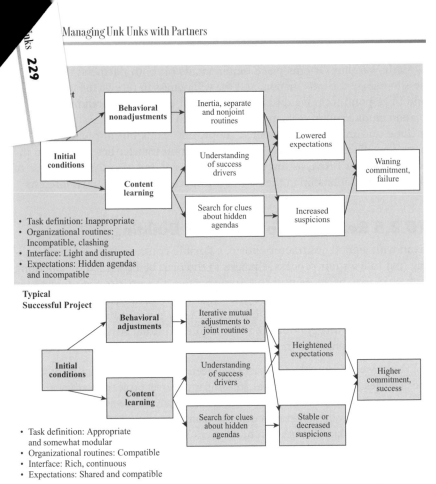

Figure 10.5 Path difference between successful and failed projects[41]

Interestingly, the above-mentioned study found that the key difference between successful and unsuccessful projects was not so much in the content learning. Rather, the behavioral adjustments influenced the effect and interpretation of the content learning. If the partner did not adjust its behavior, suspicion about its motives increased on the other side and expectations of success suffered. As an end result, this became a self-reinforcing cycle, and the project failed. If, however, the partner communicated commitment and a constructive attitude by adjusting its actions to the changing situation, information was interpreted positively, suspicion decreased, and expectations of success grew, again leading to a self-reinforcing cycle.

In other words, the *behavioral signals colored the interpretation of learning* on the part of the other party. This is, of course, related to the fair process discussion in Section 10.2.3; it is extended to a process of repeated fair process and collective action, which builds a positive (or negative) spiral of fairness, commitment, and trust. The relationship with the partner must be dynamically managed over the course of the project in order to withstand the stress of unexpected events and major modifications of the plan. Repeated cycles of mutual adjustment build personal commitment and trust, which then form the basis on which the parties can engage in the necessary collaborative problem solving when an unexpected crisis emerges.

10.3 Summary: A Process of Partner Relationship Management

We have argued in this chapter that the owner of a novel project cannot possibly hope to impose the correct actions on outside partners or contractors by contractual means alone. Of course, a powerful client can force a detailed contract down a supplier's throat. But the supplier, even if not very powerful, can usually find ways to block or retaliate, especially if emerging unk unks force changes in the plan and require the supplier to contribute to new problem solving.

We are not arguing that contracts should be done away with. On the contrary, we argue that they must be shaped in a certain way that encourages constructive behavior, and that they must be complemented by supporting measures of trust and commitment building. Figure 10.6 summarizes these complementing measures. The first step concerns partner choice. In novel projects, contractors should be chosen not based on price but on two different criteria: first, what they contribute to the competence bundle that is necessary to cover the eras of looming unk unks, and second, the "chemistry," the initial conditions of compatibility that influence the subsequent chances of constructive collaboration.

The second step refers to the contract itself. The contract should set out clear responsibilities and allocate risks and rewards in a way that is commensurate with the respective parties' abilities to handle the risks. However, the responsibilities should not be set out in the form of detailed task descriptions (because tasks will certainly change when unk unks emerge), but in terms of contributions to the general mission of the project. The contract must be flexible in the details, to allow for change. Finally, the contract should specify regular ongoing collaborative activities, to ensure shared problem solving (although this is usually called "dispute resolution mechanisms," in novel projects, collaborative activity must become routine and normal, not triggered only by a dispute).

The third step is concerned with fair process. In a novel project, partners will almost certainly have to swallow undesirable changes of the plan or perform unwanted extra activities at one point or another during the project. It is human nature to be willing to accept such outcomes only if the process is fair—that is, if their opinions and objections are heard, the changes are transparent and clearly explained, and there is no suspicion of hidden agendas. Following fair process helps to prevent anger and blockage of change.

Finally, fair process must be embedded in an ongoing building of relationships. Mutual adjustment, the willingness to go the extra mile and change one's own way of operating in order to facilitate cooperation with the partner, shapes the interpretation of events as they unfold, enhances trust in the partner's motives, and creates an attitude to helping out. If the parties successfully establish mutual expectations of collaboration and performance, a positive self-reinforcing cycle can arise that helps the parties to work through the inevitable unk-unk-caused crisis.

In summary, constructive behavior in a threatening situation, when uncontrollable events seem to take over, must be built on a *web of mechanisms* that prompt the partners to resist becoming opportunistic or abandoning the project. None of the steps alone is enough; specifically, contracts alone are not enough.

The CEO of a large engineering company expressed his intuition to us as follows: "To successfully collaborate with partners to not act opportunistically in the short term, but to be willing to contribute to the long-term goal of the project, you should do the following: (1) Create strong brand identification, (2) show your long-term game plan and create buy-in, (3) repeatedly articulate the long-term goals, (4) create emotional equity in the project, and (5) have your partners participate in fashioning the vision. And whatever you do, never compromise your credibility." In other words, he was saying that a successful collaboration requires a common interest (in this case, in a brand that all benefit from; this point reflects this company's situation and is not generally transferable), being credible and transparent, and managing a positive relationship that produces positive emotional energy. This CEO's intuition is consistent with our process in Figure 10.6, which is a bit broader and more systematic.

Even this web of motivating mechanisms has a limit, of course. If the conflicts of interest become too great, for example, if unexpected price changes disrupt the economics of the project such that one partner will inevitably lose money, the project may still fail. Highly novel projects are difficult. There is no panacea. The steps of engagement illustrated in Figure 10.5 will at least improve the chances of overcoming adverse surprises and achieving project success.

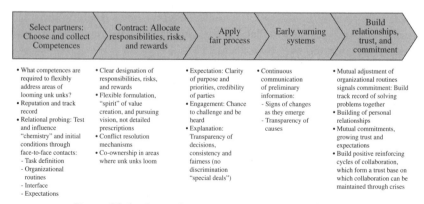

Figure 10.6 *Steps of partner management in novel projects*

Endnotes

1. See, for example, Floricel and Miller, 2001.

2. Indeed, a recent study of private-public partnerships concludes that they are appropriate if there is a combination of industry-specific competences required and public benefits, *and* a high level of uncertainty. Otherwise, a government could undertake the effort alone, or simply subcontract part of the work. See Rangan et al. 2005.

3. Contracts often have some aspect of "hierarchy," i.e., management ability to decide on the spot what needs to be done in a given situation, but the scope of such open-ended activities is usually very limited; see our discussion in Section 3.3.2, and see Stinchcombe and Heimer 1985.

4. Some authors call this a shift from contracts to "governability" in the face of unforeseen changes. For example, see Floricel and Miller 2001, and Miller and Lessard 2000.

5. This section is based on public sources, specifically, Smith and Walter 1989, *Economist* 1989, Palmer 1990, Channel Tunnel Special Report 1990, O'Connor 1993, Genus 1997a and 1997b.

6. In 1978, tunneling had actually begun when the project fell through, and the project that began in 1987 started as an existing hole in the ground on the English side at the Folkstone chalk cliffs.

7. Source: Channel Tunnel Special Report 1990, p. 31.

8. For the explanation of residual uncertainty, see Chapter 1.

9. The 1987 revenue forecasts for 2003 (10 years after planned start of commercial operations) were £642 million (see Smith and Walter 1989); actual 2003 revenues were £584 million. Not only are the total revenues well predicted but also their composition, with a bit more error. Eurotunnel's big problem lies in interest payments of £318 million (in 2003) on its debt, which makes it impossible to turn a profit.

10. This is further explained in VF 1989, which is also the source of the figure.

11. For example, Genus 1997a and 1997b.

12. To be precise, the predecessor entity of TML won the bid, CTG-FM, a consortium of contractors and banks. Then, TML, the contractors by themselves, split off in order to bid for and be awarded the construction contract.

13. Genus 1997b, p. 183.

14. The first round was taken up by the construction consortium and the promoting banks at the outset. A third equity tranche came from a public offering in 1987, after the stock market crash. It raised £770 million. A fourth public offering of £560 million was placed in 1990, after the crisis with the banks was resolved. It was a requirement that the project raise above 20 percent of total financing in equity.

15. Genus 1997b, p. 184.

16. The name of the project is disguised. It is cited from Floricel and Miller 2001.

17. The technically minded reader may note the similarity with the Circored project. There, the iron ore circulated in order to maximize reaction surface of the chemical reduction. At South Trunk, burning gases circulated to, again, maximize reaction surface and ensure efficient and complete burning. The two technologies are closely related, indeed: Lurgi has a different business unit that builds CFB power plants.

18. See Dohmen et al. 2004, Ulrich 2003.

19. Several more examples are cited in von Branconi and Loch 2004.

20. Greenburg 1975, cited from McDonald and Evans 1998, pp. 1–2.

21. Zack 1996, p. 29.

22. Floricel and Miller 2001, p. 448.

23. This is consistent with what Miller and Lessard 2000, and Floricel and Miller 2001, call "project governability."

24. The *Economist* (2004) offered this to-the-point formulation in the context of airport construction projects.

25. Floricel and Miller 2001, p. 449.

26. These initial conditions were identified, in the context of the management of alliances, by Doz 1996.

27. See Lasserre 2003.

28. Zack 1996, p. 29.

29. DeMarco 1997. This is a *novel* about project management, including contractual issues. It is fun to read and embodies the practical knowledge of an experienced professional.

30. See Floricel and Lampel 1998. The difficulty with this study is that the differentiation between behavior-based and outcome-based contracts is very indirect; the available data did not allow the authors to distinguish what types of behavioral conditions were actually incorporated in the contracts. The findings of the study are consistent with "agency theory" from economics.

31. See Genus 1997a, p. 421, and Stinchcombe and Heimer 1985.

32. See Simon 1951. Simon makes the argument that such flexible contracts overcome incentive problems in situations of uncertainty, and Loch and Sommer 2005 show that activity-based (as opposed to outcome-based) contracts help to encourage constructive behavior, provided that the owner can monitor what the contractor does.

33. This has become common practice in the automotive industry. Many car components today are codeveloped with suppliers, or the innovation even stems from the supplier. Supplier engineers work alongside the development engineers in many car manufacturers. The suppliers are responsible for prototyping and testing, and if changes occur (for example, because of change somewhere else in the car, or because of competitive responses in the car's design), the suppliers are reimbursed the costs.

34. Stinchcombe and Heimer 1985, p. 126.

35. The "cheating module" was established by psychologists; see Cosmides and Tooby 1989 and 1992, and Gigerenzer 1993. The empirical evidence supports the theory in biology (cf. Trivers 1971) that we, as humans, should have emotional mechanisms enforcing social cooperation because we are a social species and depend on one another in the group.

36. This is taken from Kim and Mauborgne 1997.

37. See Kim and Mauborgne 1991 and 1995.

38. This is recounted in Bakshi 1995.

39. *Economist* 2004, and personal discussions with managers.

40. See Doz 1996.

41. Source: Doz 1996, p. 75. The author uses the word "project" in describing the alliances, consistent with our view of a significant overlap between alliance management and project management.

Managing
Stakeholders

After discussing the internal management infrastructure (Chapter 9) and the management of external partners (Chapter 10) in novel projects, we must address one more set of parties that are important: project stakeholders. Stakeholders are parties who are not participants in the project (as opposed to partners), but who are affected by it, have an interest in it, and can influence it. Overlooked stakeholders can bring a project down, although they may not have official power. We start with an example and then draw the lessons.

11.1 The Project of the Flying Car[1]

One gray winter morning in late February 2000, one Mr. Finisterre stood on the doorstep of Frédéric Normand, the "idea scouting manager" for external innovative ideas at Lemond Automobile SA, a large European car manufacturer. Finisterre was a thinker and private inventor, and had brought along drawings of his idea: a flying vehicle. Antoine Alsace, the Innovation Department manager and Normand's boss, happened to pass by and saw the pictures. Finisterre's idea connected to something he had long (although unconsciously) been looking for; he was hooked right away. "Imagine you're in a gigantic traffic jam, and you put your wings on and simply fly over the traffic jam! We ought to do something like that!"

11.1.1 Concept Generation—Three Ideas Emerge

Normand organized a kickoff workshop in March, to which he invited Olivier LeMans, from the New Car Concepts Department, and Philippe Ardeche, a senior engine design manager for the high-end model range. Both were known to Alsace as particularly innovative and as flight enthusiasts. The team quickly named itself "Vol de Nuit," in remembrance of the French pilot hero, Antoine de Saint-Exupéry.

Ardeche brought several articles to the workshop, showing that the idea was far from new: It had been pursued for the first time in 1917. Since then, amateur designers had tried, and sometimes also succeeded, in building prototypes of flying cars. However, no one had succeeded in building anything that combined the full capabilities of a ground vehicle and an aircraft together into a single vehicle, nor had full-scale development been attempted. In the discussion, it quickly became clear that Finisterre's propeller-driven concept was impractical as a ground vehicle ("imagine pedestrians ducking for cover as it blows up a dust storm!").

Ardeche brought the discussion down to earth with the realistic comment, "Maybe we shouldn't start by trying to make a sport utility vehicle fly . . . let's proceed in small steps." This sparked an idea in LeMans. From weekly department meetings, he knew that at the time, his colleague, Jean-Pierre Breton, was working on a three-wheel curve-leaning experimental vehicle, with a shape that resembled a sailplane. Ardeche suggested building a flying motorcycle. He would talk to his long-time friend, Roussel, an ex-professional sport pilot who now had his own ultralight airplane company, and was known throughout Europe.

That evening, LeMans asked his colleague, Breton, for the three-wheeler's package plan. He worked all night and produced the first concept drawings for the "DuoSport," which he brought along to the next meeting. At this point, Breton joined the project team, to be able to consider the DuoSport's needs in the development of the ground vehicle.

In addition to Alsace, Normand, LeMans, Ardeche, and Breton, the emerging team also included Gérard Picardie, the original designer of the narrow-lane concept that Breton was now (10 years later) building. Picardie was now design manager for ergonomics and became a valued advisor in the team. André Simon was a controller with an open mind for innovative ideas, who came on board to help the team gain access to all available channels of funding and to control costs. Finally, Christelle Labelge, a business student from Metz, joined the project team while writing her *maîtrise* thesis, filling the "project office" role for the emerging team. The organization was loose and had no formal project manager.

A month later, Ardeche mentioned the idea of building a hybrid ground-air vehicle to Roussel at a local air show in Nancy. Roussel responded with enthusiasm: "Philippe, guess what? I have been dreaming about a flying motorbike for years!" Right there in Roussel's exhibition booth, they made the first concept drawing for the "FlyBike," a standard motorcycle combined with a Delta wing (Figure 11.1, top). This would emerge as concept 2.

At the same air show, Labelge carried out a small, informal survey to sound out potential interest in such a crossover vehicle. The survey indicated general interest, although the numbers were highly uncertain, and for the foreseeable future, the market would, at best, be a niche.

In the summer, a third concept added itself to the stable. When Breton heard Ardeche talk about the flying motorcycle, it brought to mind another of his ongoing projects, the "Leonardo Sport." Lemond and the Italian motorcycle company Aprilia had formed a marketing collaboration to appeal to young urban consumers, and as part of a mobility service concept, Breton worked with Aprilia to design a slimmed-down version of the Aprilia Leonardo ST 150 scooter. At 100 kg, the Leonardo Sport was only half the weight of the Aprilia SL 1000 Falco touring bike that Ardeche wanted to use, so they should be able to get that to fly! He had built an exploratory prototype, and he still had a second set of parts "in reserve." With LeMans and the company Plastic Omnium, the external partner with whom he had worked on the narrow-lane vehicle, he explored the concept of adding a foldable wing to the Leonardo Sport. When they were sure they could do this for about €90,000, they dubbed it "SkyScooter" and proposed it to Alsace's astonished team (Figure 11.1, bottom).

11.1.2 From Concept to Reality

This was how far they could go without a budget. Now, serious work had to begin, requiring resources and, thus, the support from higher up. Thierry LeCorse, the executive vice president for research and advanced development, found the idea of a flying vehicle exciting. In a confidential meeting, he commented, "I've always wanted to see a really spectacular idea. Most things we do are so incremental. We need a visionary project for a change."

With limited resources and lacking the full breadth of necessary know-how, the team quickly decided to utilize external partners as much as possible. They persuaded Plastic Omnium, already the external partner for the narrow-lane concept, to act as general contractor for the DuoSport. This eliminated the need for complex contract negotiations (including secrecy and intellectual property rights). In fact, Lemond's involvement did not even have to be revealed at an early stage. Plastic Omnium hired a project manager, Sébastien Savoie, just for this project, a level of support that Alsace had not even hoped for at the outset. They also coordinated the SkyScooter project, subcontracting execution to Marchin & Son, a small "motorized power sailplane" company and flying school. Marchin Senior was widely known as the "father of ultralight airplanes" in France.

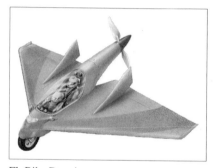

FlyBike Drawing

FlyBike Mock-Up

Sky Scooter Functional Prototype

Figure 11.1 The discontinued selectionist trials: FlyBike and SkyScooter[2]

The advantage of having an established contractor became evident when the contract negotiations for the FlyBike between Roussel and Lemond's legal department dragged on for months. The contract was signed only in early October, months after the project was initialized. Only then did Roussel receive the base motorcycle, which Ardeche had secured from Aprilia, and could finally start construction. At that time, the feasibility study and construction calculations for the DuoSport, performed by the engineering company Pinson Engineering, were complete.

11.1.3 Prototype Execution

On December 10, the team held the first workshop for everyone involved. In this workshop, the three subproject teams found out for the first time that there were three parallel efforts, and each subteam presented their status. Alsace had also commissioned a short film showing photos of the three prototypes and featuring interviews with the participants. The film was intended as a powerful communication tool to sell the project within the company.

The external partners were excited to work with Lemond—the Pinson Engineering people had first dismissed the idea of a flying sport vehicle as crazy and came on board only when they heard that Lemond was behind it. But several of the external partners were upset that they had not been told about the concept competition. Because of the contract delay, Roussel was lagging behind, but in the two and a half months, he had nevertheless managed to produce a set of animated drawings and a plastic model of the outer skin put on top of the SL 1000 Falco touring bike.

On January 29, 2001, the SkyScooter actually flew for the first time around a small airfield outside Montargis, with Marchin Junior as its pilot. In February, the 1:6 model of the DuoSport flew for the first time. This was actually fraught with difficulty, and it crashed several times because it was too small to maintain a stable airflow on the wings. The 1:2.5 model flew perfectly at the first attempt in August of the same year. Progress was incredible because everyone involved had their heart and soul in the project and worked day and night, even without pay. Several of the external partners reduced their engineering hour rates substantially, and everyone worked much more than they billed for. The Lemond internal team was essentially doing this project "on top" of their regular jobs, anyway.

At the end of April, the project was advanced enough to be presented to Michel Loiret, the CTO and head of engineering. The team presented 1:1 mock-ups of the three concepts, with photos of the successful flights, next to each other. During the presentation, LeCorse was rather tense. While initially enthusiastic about the idea, he had not expected the project to advance so quickly. This, he thought, was getting out of control—where would it end? But Loiret was excited: "In the next presentation of new vehicle concepts to the CEO, why don't we fly the SkyScooter over the heads of the group?"

Unfortunately, the excitement did not last long. In the next team meeting, resource problems became pressing. Through various channels, they had cobbled together a total budget of 1.9 million for all three concepts. The bulk of the money would have to be spent on the fully functional prototypes, but it now became clear that the budget did not suffice and no more money would be forthcoming. The team would face hard choices on which concepts to keep and which to discontinue. The SkyScooter was the first to go. In April, it was announced that the mobility service concept with Aprilia would be terminated. This meant an end to the development of the Leonardo Sport, which was to have served as the base vehicle for the SkyScooter.

By mid-May, money was still short and the team decided to stop the development of the FlyBike as well. It was not as advanced as the DuoSport, and Roussel's effort had been disappointing in the presentation in April, lacking both construction and design progress. In addition, the FlyBike was seen as more risky because it required a fundamental reconstruction of the base vehicle (the SL 1000 Falco was much too heavy and needed a lightweight composite material frame that had to be developed from scratch). Ardeche felt that "his" project, the FlyBike, was disadvantaged because less time and money had been invested in it than in the DuoSport. But he acquiesced because he, too, had been disappointed by the progress. Roussel agreed to make another presentation in October, at his own expense, in the hope of reconsideration.

Development of the DuoSport progressed on schedule. The bigger 1:2.5 model, which was a better predictor of the flying properties of the future prototype, flew successfully in August 2001, a little over a year after project start. The final prototype would incorporate advanced lightweight carbon fiber materials and fly-by-wire technology, both of which would be directly transferable into Lemond's mainstream car development, after successful implementation in the project. Figure 11.2 shows a (disguised) picture of the full-size DuoSport prototype.

11.1.4 Friction within the Development Team

While the project made breathtaking progress during the 18 months of its development, strong personalities clashed. Early on, rivalry developed between the champions of the two main concepts, who were both convinced that they were right. Ardeche was more senior, and the widely acknowledged expert, whenever the topic of flying came up. In his view, the younger LeMans had no business coming up with a competing proposal. LeMans felt unfairly treated because Ardeche's bias against LeMans's concept, expressed at the level of Ardeche's peers, including LeMans's former boss, possibly contributed to a lower annual evaluation for LeMans. This conflict led to some bad blood within the team.

Communication with upper management also turned out to be more difficult than expected. Jacques Ardennes took over the position as Alsace's boss in March of 2001, long after the project's start. Progress was incredibly fast, so Ardennes had to face the possibility of being confronted with a fait accompli. Thus, he was initially cautious, and the team suspected that he might not be on their side, fearing for his career if they advanced too far without the go-ahead from the top. Thus, communication with him was uneasy and caused uncertainty on both sides. It turned out that he was looking for other ways to bring the project into the company's official funding system, thus avoiding a complete halt to the project.

Figure 11.2 The DuoSport prototype[3]

11.1.5. Friction among External Partners

Major irritations arose with the external partners when the two concepts were stopped. Marchin had started to work essentially based on trust, without official intellectual property rights, believing that he would be able to continue to develop the SkyScooter without paying royalties. He shortened testing delays by signing a piece of paper before the maiden flight, to the effect that "the prototype was his and that his son would fly at his own risk" (had Lemond processes prevailed, the insurance question would have taken two months). The decision to stop the project, which to him seemed out of the blue, disappointed him. Moreover, he was hurt that Lemond had filed the patent without him, although he had been promised a very cheap license to commercialize the SkyScooter, if he wanted. He concluded, "This was the last time I worked with a big company."

Roussel was even angrier. He was already upset when he found out in December 2000, eight months after starting development on the FlyBike, that he was competing with other concepts. He fumed when he had to wait outside during the presentation to the CTO in April 2001 while they were discussing his baby without him being there to defend it. When the FlyBike was discontinued shortly afterward, his interpretation was that the decision had already been made at the presentation.

These irritations did immediate damage to the morale of the Plastic Omnium people, who wondered whether they, too, might be tossed out at some point. Moreover, Roussel and Marchin could destroy Lemond's reputation as a reliable partner in the small and clubby flying community, possibly compromising Lemond's ability to revitalize its efforts in the future.

During preparations for the maiden flight, protracted technical difficulties recurred, each one of them small, but collectively, they caused a delay until December 17, 2002. When the maiden flight finally took place, it was very successful; in fact, the test pilot undertook several flights that day. This prompted exhilaration, but at the same time, the event was marred by Le Mans and Breton's anger and frustration because of Breton's negative annual evaluation. Everyone who saw the DuoSport reacted first with incredulity and then with enthusiasm. And yet the project had run into political traps and resistance in the wider organization.

11.1.6 Selling the Project to the Organization

The Vol de Nuit project had several arguments suggesting a strong strategic rationale. While hybrid ground-air vehicles would clearly be, at best, a niche market for the foreseeable future, there were signs that a market for personalized air transport was emerging, both on the customer and on the technology side. For example, wealthy people in São Paulo or Monaco now bought a helicopter rather than a Ferrari—São Paulo had the highest helicopter density in the world. In the United States, small airplanes were already widely used in the Midwest and Texas, and a hybrid vehicle might spark wide interest. There were also several independent efforts reported in the press of developing much-lower-cost small airplanes. It was expected that these trends would significantly widen the market for individually owned or used air vehicles.

Lemond's competition also seemed to recognize the idea that the third dimension, air space, might gain importance for them. Toyota was currently working on a business airplane, and Audi had even approached Roussel, as a Europe-wide known expert, to ask him about a three-dimensional mobility idea that was similar to Vol de Nuit. However, he maintained a silence because of his contract with Lemond. NASA had presented a study on Dual-Mode Air-Car Concepts at the AirVenture 2001 in Oshkosh, where the large potential for such vehicles in the United States was stressed: Apart from the 29 major airports, accounting for 75 percent of air traffic, there were thousands of smaller ones. Ninety-eight percent of the U.S. population lived within 20 miles of at least one public airport.

In addition to the creation of a new market niche, which, although very risky, could be huge in the long run, the Vol de Nuit project offered several different strategic benefits to Lemond. First, there was huge PR value in being the first to credibly develop this breakthrough concept (which had impressed everyone who had seen it). Second, regardless of whether the DuoSport would ever enter the market, a technology transfer into car development was virtually guaranteed: The DuoSport design was optimized for carbon fibers. The transfer of these lightweight materials into the car had been slow because of the need for different design principles, and the knowledge gained from the Vol de Nuit project could be transferred.[4] Third, the DuoSport incorporated fly-by-wire technology, which was already recognized as important for cars in the future (eliminating wire harnesses and pipes). Fourth, the DuoSport tested a new graphical man-machine interface, which allowed instant switching between car mode controls and flight controls, and offered 3-D graphical steering and orientation support.

Finally, Vol de Nuit fit Lemond's newly announced technology strategy. In September 2001, the CTO made a presentation to the managers of the engineering division, stating, "We must move from being a leader in accessories to being the core technology leader. The technical substance of a product will become the most important differentiating factor. We must learn to achieve at least one major technical innovation per year." The CTO also stressed the importance of cooperating with external partners, an approach successfully used in the Vol de Nuit project. He also urged a change in management style "that continuously looks out for new ideas and motivates their team members to do likewise."

On the other hand, the company was entering a period of lower profits after several years of expansion, and the entire organization was under pressure to cut costs. There was less patience for "far-out" concepts. The project also faced skepticism from other departments. Marketing saw the whole thing as a distraction. The chief designer had been given the mission to establish a "common recognizable design language" for Lemond, and thus he saw this project as an unauthorized design effort that should have gone through him.

The supportive stance of upper research management began to reverse. The research and advanced development group had overrun its budget that year. On the defensive, and facing resistance from the side and from above, they became worried about the reaction of others who said, "There is no money left to get all our new car introductions ready on time, and you have money for something like *that?*" Skepticism became prevalent, although the total cost of Vol de Nuit was low.

A decisive blow came when the External Communications department joined the act. They proceeded without further coordination with Alsace, calling several newspapers to reserve a full page for "a big announcement" in June 2001. Upper R&D management heard about this plan, and feeling bypassed and overrun by events, they slammed on the brakes. In a last-minute effort, the letters to the newspapers, containing the press release, had to be hand-picked from the outgoing mail baskets. The project team ended up being blamed for not keeping upper management properly informed. On the following Monday, all communication was called off.

The team members became very frustrated and, at the same time, ever more determined to make the DuoSport fly. To give the reader an impression of the personal initiative and risks that the team members were willing to take, Box 11.1 presents excerpts of an April 2002 interview with Jean-Pierre Breton, core team member and original developer of the three-wheel fun-sport vehicle.

In spite of its technical success, DuoSport development was barely allowed to continue until the successful maiden flight (it had to be kept secret even from other departments, in order to protect it), and then it had to be shelved. In a last-ditch effort to drum up support, a team member visited a senior marketing manager in order to explain the potential of the project in communicating the brand. The marketing manager greeted the delegation with the words, "Ah, you're coming for the DuoSport. I haven't seen it, but I have heard of it. Well, we can make this short. This doesn't fit our brand." He drew a diagram like the one in Figure 11.3.

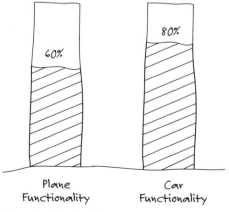

Figure 11.3 The final verdict on the DuoSport.

...ns of a Design Engineer.

...g into quite a bit of headwind even with my three-wheel fun
...cause I contracted with an external freelance designer after our
...er refused to work on it. People get crucified around here for going
...the head designer, and he has already written memos to all department
heads to block any presentation of my prototype (with the excuse that the
engine is supposed to be presented to the public only in 2005).

For the survival of the DuoSport, I have now started to lie and to hide info from
other departments, to prevent them from killing it. Last week, I made a stupid
mistake; I went to our in-house insurance agency for the [legally required]
prototype testing insurance. So, I call them, and they send me an official letter
(I hate official letters!) admonishing me about the risks and telling me this must
be coordinated with legal, and that they need all legal contracts with our
external partner who builds the physical prototype. So I dutifully send the
documents, and the next thing I know is they want a detailed project report
with all background and history. At this point, I realize I have made a mistake.
I stall for time and schedule a meeting, but I tell the caseworker he would not
get a written report from me, so he cancels the meeting.

OK, so I went to my boss and told him what happened. He asked whether
there wasn't an alternative insurance available outside, as our in-house
colleagues didn't seem to like our project too much. I say, sure, I get the
insurance outside! Half an hour later, I pass by the in-house insurance
department and get all my documents back (they looked pretty stumped!).
The rest was easy: a call to AXA; I fax a one-page risk description to them, one
more phone call to clarify questions, and the following Monday my boss signs
an airplane owner and operator insurance policy. We send it to the FAA, and
one week later we have our official permit for the prototype tests.

But now it's getting interesting: Just today, I received an official letter from
legal, warning me to not buy insurance outside, and ending like this: "We
hereby send you in writing the demand to send us complete information about
your project." I immediately called the caseworker and said there wouldn't be
any tests, and therefore no insurance and no need for their services. Of course,
he didn't believe me and said he would send me one last official demand
before taking additional action. I had to control myself to not tell him he
could kiss my . . . . I said, "Do what you have to," and hung up.

But the permit is the last barrier before we fly this baby in June. The DuoSport
will fly, you can bet on that. If the company refuses to pay the insurance, well
fine, my colleague and I pay the €880 ourselves, we can just about afford that.
I'll keep my boss out of this to protect him. That's the only way to get the
DuoSport to fly, without armies of bureaucrats and know-it-alls from corporate
running our project into the ground.

"See, with a vehicle like that, there's got to be compromises between
driving and flying. Let's say it has 80 percent of the functionality of a fun
sports car. And as the plane part is new, perhaps 60 percent of the functionality of an ultralight plane. What does that add up to? A compromise.
But we are promising our customers creative cars that work. We can't make
such a compromise. We can't do this."

Support in the organization never materialized. The successful prototype was moved into the basement of the technology center, where it sits at the time of writing this book.

11.1.7 Management Systems for Selectionist Trials in Vol de Nuit

Choice of Selectionist Trials

The Vol de Nuit project clearly faced unk unks. The construction of a flying car had been tried before, but never on a professional level, with the goal of commercialization. While there was some evidence that a market was emerging for personalized air transport, both the time horizon of this happening and the requirements of future customers were completely unknown, and would have to evolve with the products offered. This left plenty of room for unk unks to arise from the market side. In addition, some of the concepts involved fairly new technologies. For example, the DuoSport intended to use carbon fibers, fly-by-wire technology, and a new 3-D graphical human-machine interface. There were no fundamental unk unks (gaps of knowledge)—all technologies used were understood in principle and had been used before elsewhere. Thus, the team was certain that it would be able to make all three concept prototypes fly, if necessary—it was only a question of time and cost. In that sense, the technical uncertainty represented (significant) variation and risks. Still, this made the outcome of the venture a lot more uncertain, and the detailed nature of the problems that would arise was unknown.

Indeed, several unforeseen problems *did* arise during the project, even during the period before the manned maiden flight (which finally took place in December 2002), long before market introduction. The French Aviation Authority, for instance, required documentation of the programming of the steering software, something the team did not foresee, which had a significant impact on the project, both in terms of costs (an additional person was needed for the documentation) and of time. On the technical side, the prototype of the DuoSport experienced heat management problems with the engine, which delayed the maiden flight by over three months. Furthermore, the personal conflicts in the team and the resistance in the organization were unanticipated by the team, inexperienced in managing such projects. And the real source of unk unks, the reaction by the public and by the market, was still to come.

The Vol de Nuit team chose to pursue three selectionist trials. This choice was not entirely conscious—three ideas simply came up. However, they were also cheap, so according to our decision tool discussed in Chapter 7, the parallel trials made sense. The timing of selection (of the DuoSport, rather than the FlyBike and the SkyScooter) was not planned but imposed by budgetary constraints. No market tests had yet been performed; the choice of the DuoSport was made based on the team's judgment, taking into account technical risks (which were estimated to be very high for the FlyBike because a new, lighter motorcycle would have to be developed from scratch) and the team's feeling of market potential (the SkyScooter was judged to be able to serve only a small niche segment, that of leisurely "skywalking" at low speeds).

As the choice was forced by limited funds, it could be based only on preliminary information. The tests were highly imperfect (as we discussed above). They did not reveal any information about unk unks in the market, nor did they provide perfect information about technical feasibility. However, they *did* provide enough technical information to allow the project team a judgment of technical riskiness. Although this judgment could not be proven, the team felt comfortable about the choice (even the champion of the FlyBike, Philippe Ardeche, agreed, in the end, that the FlyBike was a long shot). One can argue that the Vol de Nuit project team tried an appropriately small number of concepts, and quickly and cheaply reached the point of eliminating two of them. And yet, given the preliminary information base on which the choice was made, one might also argue that they could have made the choice based on drawings only, even earlier, had the egos of the competing designers not gotten in the way.

Management Systems
Connecting back to the framework of Chapter 9, we can summarize the management systems used by the Vol de Nuit team as follows.

Planning System. A shared vision clearly existed: a vehicle that would get the owner around on the ground and, for longer distances (or rising over traffic jams), through the air, with fun. In fact, this was so exciting that the internal team members worked on this project in their free time, and the contractors were willing to work for reduced fees. The intermediate diagnosis criteria were judgments on technical risks and market potential.

Monitoring System. The team diligently used mock-up and prototyping cycles to evolve each concept in small steps that could be judged. The prototypes were evaluated based on visual (aesthetic) performance and recognizable technical risks. Experimental cycles were only two to three months long.

Coordination System. Coordination took place only at the level of the project leadership and the internal core team. The stopping criteria were relative: With the elegance and the progress of the DuoSport, the other two concepts looked progressively less promising (the FlyBike because of technical risks and lack of progress, and the SkyScooter because of market potential that was judged much more limited than the DuoSport). This is an important principle: The progress of one concept offers information about the performance of the others.

Information System. Information was *not* shared across the parallel projects because the core team feared that explicit competition would endanger the enthusiasm and energy of the partners. This caused problems—the external teams *were* frustrated when the parallel projects were revealed, and some useful technical ideas might have been worth sharing.

Evaluation and Incentives System. There were no success premiums; de facto, process incentives were used (the contractors were paid for effort that was judged diligent). However, there was a de facto winner, namely, the team that was allowed to continue. In fact, Marchin & Son wanted to pursue the SkyScooter on its own, with its own money (it would have needed to sell only a tiny number of planes in order to break even; it was not in this for the money, anyway). Marchin & Son was blocked by Lemond's legal department, which caused major irritations and cost Lemond some external good will.

In summary, the Vol de Nuit team managed the selectionist trials well, in that they explicitly pursued several approaches and eliminated two of them relatively quickly. They could have done better in the sharing among the parallel teams. Sharing success was perhaps somewhat more difficult than if they had worked with internal teams rather than external partners. However, the subteams could have been informed earlier that they ran in parallel to achieve one common goal. Also, the FlyBike and DuoSport teams fiercely competed internally (and indeed, some of the tension among the teams was caused by that internal competition more than by the fact that the subcontractors were external), to the extent that the FlyBike team took it personally when their project was discontinued. While disappointment cannot be completely avoided, creating the feeling of achieving a common goal together might have eased the conflict.

Although there were some limitations in the management systems used, accomplishing a successful prototype maiden flight in less than two years, on a shoestring budget, was an impressive achievement. The biggest problems facing the Vol de Nuit project came not from technical development, but from the interaction with the rest of the organization, the stakeholders who had no official connection to the project.

11.2 How Informal Stakeholders May Hold Up a Project

Why did this innovative project of the DuoSport fail to win support and funding at Lemond even though it had many good arguments going for it? Figure 11.4 lists the dynamics that possibly contributed. At the top of the figure, we see the stakeholder behavior that may help a project: most importantly, "goodwill," or at least the absence of resistance.

In addition, stakeholders may be able to help a project team with resources or information. Stakeholder behavior is influenced at four levels, each of which may cause resistance strong enough to kill the project.[5] Often, only the top level is discussed explicitly, namely, the strategy, the hard business arguments for or against. This level is "above the waterline."

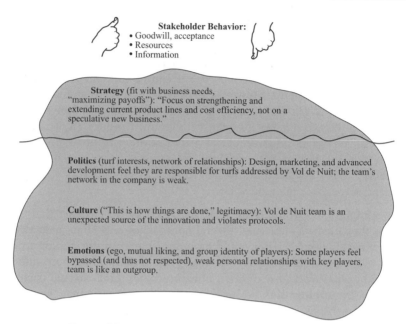

Stakeholder Behavior:
• Goodwill, acceptance
• Resources
• Information

Strategy (fit with business needs, "maximizing payoffs"): "Focus on strengthening and extending current product lines and cost efficiency, not on a speculative new business."

Politics (turf interests, network of relationships): Design, marketing, and advanced development feel they are responsible for turfs addressed by Vol de Nuit; the team's network in the company is weak.

Culture ("This is how things are done," legitimacy): Vol de Nuit team is an unexpected source of the innovation and violates protocols.

Emotions (ego, mutual liking, and group identity of players): Some players feel bypassed (and thus not respected), weak personal relationships with key players, team is like an outgroup.

Figure 11.4   Four levels of influences on a novel project

But "below the waterline" are three more levels that often play a role without ever being explicitly mentioned, or they are overlooked: the political interest constellations of the players involved; the culture of the organization, which (often implicitly and almost unconsciously) defines "how things are done around here" and punishes deviators; and the emotional reactions of the players, who hate the feeling of being diminished in their egos, of a breach of loyalty, or of supporting someone outside the "ingroup." They are often willing to put their foot down (if necessary with fake arguments) if they feel aggravated.

11.2.1 Strategy and Economic Reasoning

An organization has a legitimate interest in pursuing projects only if they support the organization's priorities. The difficulty with that lies in the fact that strategic priorities can never be "proven": They always involve judgment calls. In other words, reasonable people can disagree about what the strategy really requires, and a dialog is required among the decision makers to come to a common judgment.[6]

That did not happen for the DuoSport project: Although good strategic arguments for the project existed (they are listed in Section 11.1), these were not shared. Marketing saw this as a distraction (thinking that the current product line needed strengthening rather than starting a new business), the car-body people did not yet see the potential of carbon fibers (except for a few advanced body development designers, who weren't being heard by the mainstream of their department), and the organization at large had entered a period of emphasizing focus and cost cutting rather than risky undertakings. But the debate never really took place—for example, Marketing dismissed the project after hearing a short description of it, without ever becoming fully informed.

It is important to realize that Lemond may have been right in shelving the project. Perhaps there really were not enough resources to pursue the project further without endangering the current business and its restructuring (the pity is that, for political reasons, it was the dialog that did not take place—see below). Shelving projects is legitimate and necessary for organizations to maintain focus, and designers must realize and accept this, without taking it too personally.

It is common practice, for instance, in car companies to stage concept design competitions, in which up to a half dozen elaborate vehicle designs are developed in parallel (for example, in clay or wood, looking like the real thing), and then top management chooses the most promising one to go into engineering. This is necessary because one cannot judge a concept design from drawings beforehand; it's too complex. The designers whose concepts are not chosen are often upset for months afterwards.[7] But, in fact, all the designs are necessary to be sure of having a promising one in the end, and no one's effort is wasted (although it may feel that way). This is part of designing in a business.

11.2.2 Politics and Influence: Differing Interests and Network Structure

The strategic view of an organization claims that the organization acts like a unit—a single entity that makes decisions to maximize its success. But that is, of course, only true in special cases (for example, when the organization undergoes an existential crisis). Most of the time, an organization is a coalition of partially conflicting interests. Every department manager looks out for her career, her resources, and her influence, and there always exist multiple and shifting alliances, which can help her if she piggybacks on them, or which can destroy her if she gets in their way.

This insight has two important implications for the designer who tries to get an idea accepted: (1) Be informed about who has what interests at heart, and how your project affects the various "interest turfs;"(2) the organization is a network in which power and influence are not completely mapped by the official hierarchy. Know who is allied with whom, so you can approach key players who then do your work and convince others for you.

Interest Turfs

At Lemond, the head designer felt threatened by the project, because if it succeeded, it would diminish his monopoly on design expertise. Marketing also felt threatened because they claimed to be the experts on the judgment of market niches, and if the project succeeded, it would imply that they had overlooked something. Moreover, all other departments were interested in limiting the power of engineering and were happy to use the budget over-run in advanced engineering (where the DuoSport was located) to score a victory. Knowing the "turf" is critical in predicting where resistance will come from, and which arguments will diffuse that resistance as much as possible (for example, by letting Marketing share the credit for identifying a new and promising market).

Influence Networks

Clearly, there are some stakeholders who do not matter (who have no influence). There are situations where they should simply be ignored. As an example, in the context of preparing the infrastructure for the Sydney 2000 Olympic Games, a major project was the building of a 14-mile sewage tunnel under an affluent part of Sydney, in order to clean up Sydney Harbor in time for the games. The project team invested a great deal of effort in getting the communities affected by the tunnel on its side, in order to avoid resistance. Ultimately, however, some of the communities clung to concerns about the venting of the tunnels (even through that, in engineering terms, was not a problem), and by the time the project had progressed to a certain point, their concerns were no longer a danger. The project then sailed through the remaining objections and finished on time and slightly over budget.[8]

However, influence is often underestimated. Project teams naturally look for parties that have *direct* and *explicit* power to disrupt the project. Nevertheless, much influence is *indirect* and works *through others*. Managers (and people in general) do not make decisions in a social vacuum but look for guidance and advice from their superiors, from their peers, and often from their subordinates. When others rely on you for information and advice, you have informal power. Informal power is often dispersed in ways that are different from the official hierarchy—yes, you have to listen to your boss, but your boss may go to a peer for the kind of information that drives his or her decision making. This kind of influencing power resides in the social network structure of an organization. While a few people are "naturals," most people do not pay enough attention to the social networks. It is worthwhile to understand who is central in the network, who is always informed, and to whom others listen. If you get those people on your side, their support tends to amplify.[9] In short, some stakeholders may have very little direct power, but still matter because of their informal influence on others.

Figure 11.5 represents four different network positions of a project team. Network A represents a team with weak internal relationships (no connections) in an environment that is also unconnected. This team possesses two strengths: Internally, it has diverse perspectives, skills, and resources, and externally, it represents a "structural hole"—that is, it connects external parties that are otherwise unconnected, and it has thus the potential to gain power by acting as an information broker among them. However, team A lacks the internal relationships, the closure needed for good group cooperation and control. It may not be able to exploit its favorable position in the overall organization.

Network D represents the opposite case. The team is connected internally and externally, and the environment is also well connected around the team. In fact, the entire organization is one highly connected, cohesive group. While this cohesive group will find it easy to cooperate and be flexible, it has access to only one perspective, skill, or resource, which bears the risk of groupthink. Cohesive networks have been shown to have difficulties in making decisions in uncertain environments.

A team with the network structure B is in the strongest position. The team is internally strongly connected and therefore able to present a united front toward the outside. Moreover, the united team is an information broker, providing the only connection between the unconnected outside parties. This information brokering gives it both power in influencing those parties as well as access to diverse information and opportunities.

A team with the network structure C is in a difficult position. While it is internally cohesive (like team B), it has few links to the rest of the organization, which, in turn, is well interconnected. This structure makes it difficult for the team to tap into external information or resources, and the structure also allows the possibility for important information to reside in the network without team members getting to know about it: The team cannot control, or maybe not even monitor, what is going on around it.

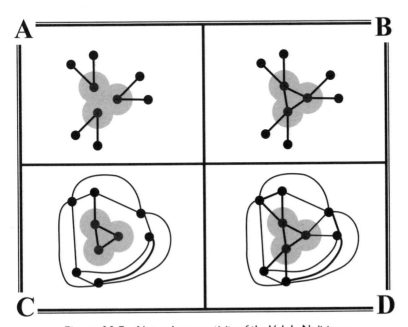

Figure 11.5 Network connectivity of the Vol de Nuit team

The Vol de Nuit team found itself in the unfavorable situation represented by case C: It had only few informal ties with the rest of the organization. The entire organization was highly connected, a cultural feature of the company that had traditionally operated based on informal ties rather than official structures. This made it difficult to use informal power to win support for the project. This unfavorable situation contributed to the fact that the team did not know exactly what the CEO had said about "submarine" (i.e., unofficial, not explicitly authorized) projects.

In summary, the Vol de Nuit team was in a very difficult situation in terms of politics and relationships. Perhaps the team could have attempted more systematically to win natural allies, possibly influential ones. For example, the head of Advanced Engineering had little informal power. The head of Manufacturing, in contrast, was a natural ally, since the carbon fiber technology was something relevant to him, and he had the reputation for being open to new ideas. Being well regarded and listened to in upper management circles, he could have had a favorable influence on the project's acceptance. However, the team did not approach him. While it had natural enemies, it did not sufficiently rally the natural allies.

The reaction of the senior marketing manager illustrates the subtle effect of network relationships. One may agree or disagree with the content of what he said. But it is striking that he made the statement without having seen the prototype or having talked to any of the team members. When people don't know you, or have not had someone whom they trust recommend you, they have no qualms about seeing you in a negative light. The network position of the Vol de Nuit team clearly played a role in the final decision.

11.2.3 Culture

Culture defines, both explicitly and implicitly, "how things are done around here." It defines appropriateness and legitimacy. The sociologist Edgar Schein[10] discovered in the 1970s that cultures are powerful organizational memories of intelligent rules—if every employee had to make a conscious decision at every turn (rather than just following a feeling of what's "naturally" legitimate and appropriate), mistakes would abound and chaos would reign. Chris Bangle, the chief designer at BMW, appeals to the culture of designing the sleekest performance cars when communicating with the organization at large.[11]

As the cultural rules are "automated" and no longer reflected upon, they carry the danger of becoming obsolete when the environment changes. This is well captured by an example in a biography of Thomas Watson, Sr., the first CEO of IBM.[12] In the 1920s, he instituted a policy that IBM salespeople should be dressed like their customers, mainly banks and large conservative companies (to fit in and to foster trust). Three CEO generations later, in the 1980s, this rule had fossilized into the famous "blue suit and yellow tie" stereotype, which made IBM salespeople look entirely out of place when they were with high-tech clients in slacks and sandals. And yet, when Lou Gerstner scrapped the blue-suit rule in 1994 (in effect going back to Watson's original philosophy), traditionalists howled that this threatened the very core of IBM's culture!

Legitimacy and appropriateness affected the DuoSport project at Lemond: Who had the right to do a "far-out" project like this? Antoine Alsace's New Business Development department had developed strategies and made presentations before but had never gone all the way to functional prototypes. In fact, Breton officially belonged to a sister department, not to Alsace's, and just worked on the project part-time. No one in the organization expected such a breakthrough firecracker to come out of this department, and it scared people. Prototypes usually came out of another department, Advanced Technology. Perhaps an "adoption" of the project by Advanced Technology might have helped (but, of course, that raises turf questions of who gets the credit!).

The second important cultural issue affecting the DuoSport was related to the path of presentation and successive authorization. The project was de facto run like a "skunk works" (industry jargon for a project running in isolation and secrecy from the rest of the organization).[13] The usual practice of the organization was to run projects relatively quickly by upper R&D management, then an investment council, and then the CEO. But the DuoSport had missed that window, having run too far ahead without the CEO being informed (partially driven by the perceived turf conflicts and resistance in the organization). The CEO was rumored to have made the remark, when seeing some pictures, "I thought we didn't do these cowboy projects any more!" The team was now trapped, not daring to show further progress to the CEO for fear of officially being forbidden to continue. Right up to the end, they hoped for the "revelation" at the maiden flight.

11.2.4 Egos and Emotions in the Approach of Individuals

Apart from strategy, political egoisms, and cultural habits, people commonly (not only in business organizations) exhibit three emotional needs that you neglect at your peril: friendship and reciprocity, group identification ("are you one of us?"), and ego. Whether or not you consider them in the way you approach decision makers or supporters may make the difference between support and indifference, between neutrality and hostility.

The first emotion, friendship and reciprocity, is a double-sided one. On the positive side, past investments in people, in the form of paying attention to them, being sympathetic, coming across as fair and reasonable, or helping out, carry benefits that can be "called in." Just put yourself in the situation of being approached with an idea by a colleague with whom you have had a positive relationship for a long time. It will be emotionally very difficult for you to tell that person that this is incompetent and inappropriate for the organization! Your natural bias will be to look for strengths in what your colleague does, to be negative only if you can't avoid it, and even then to be nice about it. On the negative side, friendship can turn into active hostility if someone feels crossed by a person who was supposed to be trusted. Friendship opens possibilities, but it also constrains you in order to keep the relationships positive.

The DuoSport team was too weakly connected with the rest of the organization to use friendship ties. In other words, the team lacked a high-level *sponsor* that could have provided the external stakeholder connections and lent his or her weight to informally influence stakeholders.[14] Upper R&D management, two levels higher, could have done so, but these managers were either cautious (because of the turf issues) or had not been sufficiently mobilized. This lack of emotional involvement made it easier for the rest of the organization to dismiss the concept.

The second emotion is the feeling of loyalty and solidarity of "Us" against "Them." In some situations, one might be able to mobilize a manager by telling him, "Look, Chrysler just presented the Dodge Tomahawk 400-horsepower motorcycle at the Detroit Auto Show, and it's just a gimmick, but they get lots of press. Do you think we should let them look more innovative than we are? We could steal their fire by showing the DuoSport!"

Third, people crave the stroking of their egos: getting credit for what they have done, receiving compliments for their competencies, being asked for their opinion, and having an influence on events. They absolutely hate the feeling of being bypassed, wrong, or insignificant. The higher they are in the hierarchy, the more pronounced the ego becomes.[15] You can harness this energy by giving someone the chance to feel significant by helping you. A humorous example was told to us by the Mexico country manager of a car company.[16] He needed to coax the Mexico City dealers into upgrading their facilities (which required a significant investment). He called them together and told them: "Only one dealer, Mr. X, is allowed to participate in our upgrading program, because this is only for the best. The others are not allowed to participate for now, and I'll keep you posted." Now the other dealers actively fought to be awarded the right to participate (and invest a lot of money) because they could not stand not to be among "the best."

11.3 Lessons: Map the Decision Influence Levels to Sell the Project

The reader may recall the definition of stakeholders: They are parties who are not participants in the project (as opposed to partners) but who are affected by it, have an interest in it, and can influence it. If a project team attempts to get stakeholder support for, or at least avoid resistance against, a project in a large organization, or in the community around the organization, the analysis in Section 11.2 helps to perform a *mapping exercise* to identify the selling points and the potential points of resistance that you are facing. Naturally, each organization is different in terms of the precise criteria at each level and in the relative emphasis placed on the levels, but the levels of decision influences are stable categories to consider. The first two parts of the mapping are about the content of the team's *arguments*.

1. *Strategy.* Understand the business priorities of the organization, and map with respect to them what your project can contribute (this may include monetary figures, or qualitative contributions, as long as you can explain them). For example, the DuoSport team at Lemond started working on a "mobility strategy," which might later convince the company to revive the project.

2. *Politics.* Map the key players, what their interests are, how the design project in question relates to each one of them (who will find it helpful, who will find it threatening or distracting?), and who the influential people are. This will imply an approach of garnering support for your project.

   Influence may be direct and explicit, for example, by the position in the hierarchy or by the control of certain key resources. If an influential stakeholder has interests that are hindered by the project, a classic approach in politics is to "trade"—the project management offers the person something in return for supporting (or not resisting) the project.

   Influence may also be indirect and embedded in a network of relationships, working through others rather than through explicit power of one's own. It is therefore important to understand the network structure, who is central in it, and who can play a *sponsor* role of establishing a connection between the project team and the central players. Drumming up informal support through a network requires *persistence*. It takes much time and effort to achieve the "critical mass" of supporters in order to swing the mood of an organization.

   The Vol de Nuit team could not rely on the officially powerful people because they had other interests or did not want to go against the general cost-cutting mood of the organization. Nor did the team have a sponsor who might have helped to rally the natural allies. In addition, the project had some natural enemies. This was very costly in terms of momentum.

Parts 3 and 4 of the mapping exercise are about the *approach* of "selling" the design idea to the organization.

3. *Culture.* What is the "appropriate" way of introducing such a project into the organization? Who are natural sources, what are the accepted channels of communication, what does authority rest on? What, in the proposed approach, feels "unnatural," and why?

   The DuoSport team was an unexpected source of a design innovation of this type, and it was trapped in communication expectations that contradicted the looming political minefield. At the same time, the company, Lemond, had a proud history of innovations and initiative-taking by teams at a low hierarchical level; the team might have appealed to other people's consciousness of that history.

Going against cultural "habits" requires, again (as in point 2), *persistence* on the part of the project team. As in convincing a user community of a new and unfamiliar design, the project approach first needs to be presented in "weak" form, until it no longer feels unnatural, before pushing with full force.

4. *Emotional needs.* What are the emotional "hot buttons" of the players and intermediaries? Again, everyone is different, but the types of hot buttons are always the same: the desire of personal loyalty, the emphasis on a common group identity against a shared outsider group, and the need of ego acknowledgment.

   This level is closely connected with the principle of *fair process* that we discussed for project partners in Section 10.2.3. If we want stakeholders to lend us their support, and to continue doing so when unk unks force unexpected changes in the project, we must inform them and keep them apprised of unexpected changes and their reasons. Otherwise, distrust will translate into a withdrawal of support.

   This level, like the third, worked against the DuoSport team: They were de facto outsiders (a general problem that skunk work teams often have) and had weak friendship ties to the network of decision makers. The levels of decision influences interacted—the team's upper management preferred to lie low for reasons of political turf, and so the pull of personal relationships was missing as a supporting force.

The DuoSport example shows how a good design idea, with solid arguments for it at the "strategy" level, failed because it was weaker at the other three decision influence levels. Mapping the levels helps you to diagnose where you are vulnerable and to devise an action plan that will maximize the former while minimizing the latter. Recognizing the four decision influence levels is the necessary preparation to navigate the jungle of influences in a large organization or the societal environment.

Endnotes

1. This section is based on Loch 2003 and Loch and Sommer 2004. The name of the project, Vol de Nuit, and the name the company, Lemond Automobiles, are disguised in order to preserve confidentiality.

2. Source: Loch and Sommer 2004.

3. Taken from Loch and Sommer 2004. The picture is disguised in order to protect the confidentiality of the design and the company. The real design is much more elegant than this disguised version.

4. Carbon fibers are very strong with respect to pull forces, but weaker than steel with respect to shear. Thus, simply replacing steel by fibers without a change in the design required much more bulk, which weakened the weight advantages and exacerbated the cost disadvantages. Changing the design in this way had to be learned over time, and the DuoSport was a first test case.

5. See Ancona et al. 1999 for the first three levels, and Loch, Yaziji, and Langen 2000 and Urda and Loch 2005 for the fourth.

6. See, for example, Loch and Tapper 2002.

7. See, for example, Bangle 2001.

8. This project is recounted in Pitsis et al. 2003.

9. For example, Ancona et al. 1999, Gladwell 2000, Baker 2000.

10. See Schein 1985.

11. Bangle 2001, ibid.

12. See Maney 2003.

13. See Rich 1994.

14. Sponsoring, or generating informal support and access to first, "bootlegged" resources, is an important function of making innovative projects happen. See Roberts and Fusfeld 1997.

15. For an illustration of how status and egos influence behavior in an organization, see Loch et al. 2000.

16. Not Lemond Automobiles but a different car company.

PART

IV

Managing the Unknown: The Role of Senior Management

The book has, so far, concentrated on the methods, tools, and mind-sets available to, and used by, the project team. However, the project team cannot accomplish the difficult task of mastering unforeseeable uncertainty alone. It is embedded in a context, and although the team has the responsibility of managing its stakeholder network (as we discussed in Chapter 11), it needs the help of senior management.

Senior management cannot simply delegate to the team the responsibility for successfully leading novel projects and then let them survive or die on their own. The task is, well, too unforeseeable and too difficult. Senior management has the responsibility of setting up the project team in a way that it is given a chance, and supporting it on the way. This is what our final chapter discusses. Chapter 12 outlines three key responsibility areas that senior management must shoulder.

The Role of Senior Management in Novel Projects

While project managers and their teams are responsible for executing projects, even in the difficult situation when unk unks are present, novel projects place a much greater demand on the involvement, time commitment, and knowledge of senior management than do routine, planned projects. Much less can be delegated, and top management must engage by investing time to be well informed, by being open to dialogue, by being willing to question its own assumptions, and by allowing the discomfort and insecurity of changing agreed-upon plans. Senior management must also be committed to *fair process* (see Section 10.2.3) in order to maintain loyalty and commitment of the project team under the pressure of unk unks.

In summary, senior management must set up an environment in which novel projects have a chance to succeed. We believe that senior management must contribute three critical elements in order to support the project team: Choose the project scope so that it fits the organization's strategy, build the organizational capabilities to deal with unk unks, and ensure appropriate sponsorship of the project. These three contributions are the topic of this chapter.

12.1 Choosing the Project Scope
12.1.1 Understanding the Strategic Rewards

Novel projects pose inherent and unavoidable risks. Even if a team manages its project impeccably, unforeseeable events may lead to failure. And, indeed, most novel or breakthrough projects go through periods of demotivation and doubt about the final success, and often they *do* fail, at least at the first attempt. This is exacerbated by the fact that large projects often require up to a third of the project in financing up front, an investment that is sunk if the project does not succeed.[1] Therefore, such projects should only be undertaken if they offer commensurate rewards.

Rewards from novelty *do* exist. Miller and Lessard state that "successful projects are not necessarily the easy ones, but those whose real value can be created by hard, creative work, leverage at the right moment and influence on the right groups."[2] Similarly, Hamel and Prahalad conclude that capturing large opportunities from new kinds of activities requires operating in unstructured areas and having the perseverance of a marathon runner.[3]

However, novelty does not guarantee strategic rewards! It is the responsibility of senior management to understand the vision of the project and to ensure that, should the project deliver, the results are compatible with what the organization wants to achieve in its business (or in a new business) and that the organization really can appropriate the results from the project. Examples abound of projects that succeeded and were then rejected internally by the organization. The Vol de Nuit project in Chapter 11 is a clear example of this.

Thus, it is the responsibility of senior management to:

- ▲ Understand the project vision and think through what the organization will do if the project succeeds.
- ▲ Clearly, openly, and frankly think through the risks that can be anticipated, assess the vulnerability to unk unks, and make these vulnerabilities known.
- ▲ Involve the key partners in this process of thinking through the rewards and risks. Of course, the project team must also be involved, in order to have access to the best possible information, and in order to create ownership by the team and a mutual understanding between the team and management.

This is costly and takes time, and moreover, it takes courage because uncomfortable findings may arise. The earlier uncomfortable findings can be articulated, the less difficult will be subsequent partner and stakeholder management (Chapters 10 and 11). If management does not perform these up-front activities, it may set the project up for failure.

12.1.2 Shaping the Project Portfolio

As highly novel projects, which are vulnerable to significant unk unks, are so risky, no organization can undertake only projects of this type. In most organizations, routine projects with well-understood rewards and risks should be the norm, and novel projects should be the exception, accounting for a small part of activity. For example, R&D organizations of large industrial companies spend only between 1 percent and 10 percent of their total R&D expenditure on risky, long-term, and potentially breakthrough work. What the right percentage is depends, of course, on one's business and one's strategy—there may be engineering contractors who carry out a large novel project that temporarily accounts for a great part of their turnover. But such contractors had better protect themselves or they could run the risk of bankruptcy.

The point is that senior management must have a very clear picture of its project portfolio and the uncertainty profiles of the projects in it. If the fraction of risky projects is high, management should articulate explicitly their benefits to provide reasons to incur those risks. Finally, the fraction of risky projects is not a given, or something that "emerges": It is a *decision*. Novelty is not fully imposed by client demands or industry competition; the design of a project includes (implicit or explicit) decisions to accept a certain level of uncertainty, or to limit the chances of being confronted by unk unks. For example, one design parameter is the choice of technology—how cutting-edge does it need to be? In electronics, we discussed in Section 10.1.2 the example of the German TollCollect project, which combined novel elements in a complex architecture. While contractual difficulties exacerbated the problem, discussions of the project also suggested that some simpler system components might have been used. In a construction project, one can choose a spectacular architectural design that requires novel techniques and thus may give rise to more unk unks, or one may prefer a more traditional architectural design and thus limit the potential for unk unks. For example, the spectacular collapse of the Paris Charles De Gaulle Airport terminal project in 2004 was due partly to its difficulty to realize its novel design. In summary, management should have a policy, or a rule of thumb, on how many novel and risky projects it allows.[4]

12.1.3 Enforce Risk Reduction in the Projects in Which Novelty Is Not Critical

The third aspect of choosing the scope is *discipline*. For those projects in the organization's portfolio for which high novelty has not been identified as critical for achieving an important contribution, novelty should be avoided. In other words, such projects should be directed toward understood customer needs and should use well-known technologies, stable processes, and a system architecture that guards against unforeseeable interactions from complexity (see the discussions in Chapters 4 and 7).

This is nontrivial and takes senior management direction because technical personnel often tend to prefer more challenging, higher-quality, and higher-performance solutions when offered the choice. This is because more sophisticated work is more interesting; it gives them respect and provides them with stories to tell within their technical community, and they may even hope to surpass the expectations of management. However, if the organization wants to avoid a proliferation of unknown risks, it must combine the flexibility of managing novel projects with the discipline of limiting routine projects to remain just that: routine.

12.2 Building Organizational Capabilities

The second key responsibility of senior management is to ensure that organizational capabilities are developed such that they enable the organization to execute novel projects. This will involve hiring, developing, and assigning the right people to the right projects, building the proper project infrastructure, and implementing appropriate project governance. We discussed these topics in Chapters 9 and 10, but they cannot be accomplished within the team alone; senior management must set the stage.

12.2.1 The Project Management Team

The less structured nature of novel projects necessitates three requirements for the profiles of the team members: experience, flexibility, and mindfulness. The first requirement, *experience*, refers not only to deep experience in the technical subject area but also to previously having witnessed unk unks and the responses to them, so that the team member does not panic or become confused when unk unks emerge. Project teams who are inexperienced in novel projects will too often come with a planning mind-set and instinctively fall back on standard PRM techniques and approaches when they are not appropriate. Building such an experience requires careful career management. For example, a large organization might groom a cadre of project managers that can cope with novel projects. Grooming these people consists of allowing them to develop through a sequence of projects in which they become more and more confronted with uncertainty and have to deal with increasingly complex external interests and stakeholder constellations. Building such networks often requires well-planned lateral career moves through different departments and geographies.

The second requirement, *flexibility*, refers to personality profiles who are not dependent on fixed routines (as opposed to people who find security only in stable work patterns) and who do not become too attached to work that has been carried out under certain assumptions, so abandoning it does not become too stressful. The project team must be able to anticipate and exploit early information if it is to benefit from early probing.

The third requirement, an *organizational mind-set*, refers to the culture and the informal, or "automatic," behaviors of the project team when dealing with novel projects. One important aspect of this culture is *mindfulness*, the ability to detect and respond to unexpected events in novel projects. As we discussed in Chapter 8, mindfulness has five components:[5]

1. *Preoccupation with failure.* Project teams must be able not only to tolerate but to seek failures, especially early in the project, and to learn quickly from these failures.

2. *Reluctance to simplify.* Project teams that are alert to unk unks try to simplify less and see more, acknowledging the complex and unpredictable nature of the project.

3. *Sensitivity to operations.* Normal operations, procedures, and processes often reveal observations that have no immediate consequence but are "free lessons" that could signify the development of unexpected events.

4. *Commitment to resilience.* A key characteristic of unk unks is that no matter how well one prepares, the unexpected *will* happen.

5. *Deference to expertise.* Decision making is pushed down in the organization, where faster detection, more knowledge at the decision-making level, and more variety in approaches increases the chance of finding a good solution.

In Chapter 8, we discussed mindfulness as a characteristic of the team that carries out the project. But, of course, the team does not operate in isolation. If senior management violates mindfulness, for example, suppresses dissent, or is not knowledgeable about key aspects of project operations, it torpedoes the team culture and risks making it difficult, if not impossible, to be open-minded and flexible.

The combination of these three requirements of experience, flexibility, and an appropriate organizational mind-set is difficult. It raises the question of whether one can find these characteristics outside the organization or whether they can be developed and sustained only internally.

It is certainly possible to hire external, experienced project teams who know how to respond to unk unks. However, the fuzziness of project targets, and the unforeseeable demands on resources, which come with the presence of unforeseeable uncertainty, require close integration of the project team with the organization. Handling unk unks happens in a particular organizational context, sensitivity to the operations requires a deep understanding of the organization, and a commitment to resilience demands a credible and respected project manager. Thus, a highly novel project often requires a longstanding intimacy with the organization. Therefore, it appears to us that it will, in the normal course of events, be difficult to hire these capabilities from the outside and have them rapidly operational.

12.2.2 The Project Infrastructure

In addition to the right people, novel projects must also have the appropriate project infrastructure. Project infrastructure includes *systems* for planning, monitoring, coordination, information management, and performance evaluation. The infrastructure must distinguish planned, selectionist, and learning aspects of projects and subprojects.

We discussed these management systems, and their various configurations, for planned, selectionist, and learning (sub) projects in detail in Chapter 9, but only from the viewpoint of the project team. However, while the team can competently *use* these systems, it has only a very limited ability to *install* them. Only senior management can install the systems of the project management infrastructure.

It is not widely accepted in organizations for multiple versions of project management systems to be needed. Most companies have an established and documented project management system, and then a "light" version of the same process for small projects or for "exploratory" projects that need more flexibility. However, while a "light" process version may indeed succeed in providing somewhat higher flexibility, this approach fails to capture the fact that a novel project, which is managed with selectionist and learning elements, needs a different approach to planning, monitoring, coordinating, and evaluating performance. As we discussed in detail in Chapter 9, what is planned and monitored is fundamentally different, concerning experiments, testing hypotheses, information sharing, and halting trials, rather than the progress toward the specified target. The systems must be concerned with a "meta level" of how a team can learn and adapt, in addition to being directly concerned with progress.

Only senior management can bring about the development and installation of systems that embody such a different philosophy. Project management, then, has the responsibility of leveraging these systems to produce value, and to feed back learning, so that they can be adapted and improved upon. The first step toward the decision to implement such systems is the understanding by senior management of what they are for, and why a different approach is needed. This is the discussion to which this book hopes to contribute.

12.3 Sponsoring Novel Projects

Project sponsorship is concerned with "behind-the-scenes" support, protection and advocacy for the project, and sometimes informal "begging" for funds (like fundraising).[6] An ideal project sponsor is a political heavyweight in the organization who can help to protect the project and to influence decision makers.

We discussed in Chapter 11 how a team should attempt to influence the network of stakeholders—the parties that are external to the project and who do not have an official role but can, nevertheless, influence the project and will do so because their interests are at stake. Again, this discussion proceeded from the team's viewpoint, although the team cannot accomplish this alone. The Vol de Nuit team in Chapter 11, for example, was effectively abandoned by its natural sponsors; its isolated position in the organization at large was only to a small degree its own doing, but reflected a failure of senior management to support the team.

Effective sponsorship is critical in novel projects, even more so than in planned projects. This is because emerging unk unks are likely to lead to team needs, such as resources, expertise, or strategic support that were not planned and are not easy to obtain. Organizations are notoriously reluctant to provide additional resources unexpectedly, on the fly, because that smells like misuse or fraud. A heavyweight sponsor in the senior management of an organization is like an additional buffer for the team, a buffer that provides some crucial flexibility in responding to unexpected events and can help fashion a solution that keeps the project vision within reach.

It is therefore the responsibility of senior management to ensure sponsorship for a novel project that has been initiated. For instance, the top management team can assign one senior manager to each of the novel projects under way. Such an assignment must, however, be "real" as opposed to just paying lip service. In other words, the sponsor should not be given this assignment simply "on top" (which implies that he or she will not have the time or make the investment to become informed and to closely follow the project and its unexpected twists and turns), but that this assignment comes with some resources or time attached.

It is important to realize that the sponsor is not only nice to have for the team. The sponsor also represents supervision by top management. Sponsors should encourage and support the team through difficult times in the project's history but should equally ask the tough questions and cool off the team when it gets too excited about unexpected progress. If the sponsor indeed stays closely involved and informed, the team has less leeway to find excuses or to misuse positive surprises to make its own life easier rather than to produce more value.

Recall that we concluded in Section 9.2.4 that "process incentives" are favored as evaluation systems to costly upside incentives if management has the ability to observe what the team does. A closely involved sponsor, while helping the team, also improves senior management's ability to do just that—namely, to observe closely what the team does. In addition, the sponsor establishes a win-win relationship with the team: "I help you, but you play it straight with me." In this relationship, the team acquires a personal obligation to perform; it goes beyond the official performance measurement criteria. In other words, the sponsor also has the ability to keep the pressure on.

12.4 Conclusion

Thus, we conclude the book. We do not have profound words of wisdom, except to express a humble hope that project managers find it useful. Project managers are our heroes. They are often underappreciated players in organizations, and yet it is they who do the novel stuff that the organization cannot accomplish in its everyday processes.

This book is meant as a resource for project teams that have to deal with novel projects. Even if everything in this book was of supreme wisdom, it would not make managing novel projects trivial. Dealing with unknown unknowns is inevitably uncomfortable and dangerous. We hope this book provides some guidance or red thread in the chaos of dealing with unexpected and hard to interpret events.

In this last chapter we hope to remind senior management that it has an important role to play. Projects do not happen in isolation but are implemented in an organizational context. This context is created by senior management. In many organizations with which we are familiar, senior management does not heed this responsibility, being too occupied with financial measures and strategy. But the recommendations in this chapter are real—if they are disregarded, the ability of the organization to successfully pull off novel projects may be severely compromised.

Endnotes

1. Miller and Lessard 2000, p. 12.

2. Miller and Lessard 2000, p. 198.

3. Hamel and Prahalad 1994, p. 37.

4. Well-known methods exist to support the risk balance of project portfolios, for example, with the "strategic bucket" model of assigning appropriate resources to new markets, new technologies, and incremental projects. See, for example, Roussel et al. 1993, Cooper et al. 1998, Kavadias and Loch 2004.

5. As we have stated in Chapter 8, this is based on Weick and Sutcliffe 2001, Chapter 1.

6. Fusfeld and Roberts 1997, p. 276.

References

Adner, R., and D. A. Levinthal. 2004. What Is Not a Real Option: Considering Boundaries for the Application of Real Options to Business Strategy. *Academy of Management Review* 29 (1): 74–85.

Allen, T. J. 1966. Studies of the Problem-Solving Process in Engineering Design. *IEEE Transactions on Engineering Management* EM-13 (2): 72–83.

Ancona, D., T. Kochan, M. Scully, J. Van Maanen, and E. Westney. 1999. Three Lenses on Organizational Analysis and Action. Module 2 in *Managing for the Future: Organizational Behavior and Process.* 2nd ed. Cincinnati: South-Western College Publishing, 1–75.

Angle, H. L. 1989. Psychology and Organizational Innovation. Chap. 5 in *Research on the Management of Innovation: The Minnesota Studies.* ed. A. H. Van de Ven, H. L. Angle, and M. S. Pole. New York: Harper & Row, 135–170.

Argote, L. 1999. *Organizational Learning: Creating, Retaining and Transferring Knowledge.* Norwell, MA: Kluwer.

Argyris, C., and D. A. Schon. 1978. *Organizational Learning: A Theory of Action Perspective.* Addison-Wesley.

Austin, R., and L. Devin. 2003. *Artful Making: What Managers Need to Know about How Artists Work.* Upper Saddle River, NJ: FT-Prentice Hall.

Baker, W. E. 2000. *Achieving Success through Social Capital.* San Francisco: Jossey-Bass.

Bakshi, A. 1995. Alliance Changes Economics of Andrew Field Development. *Offshore*, January, 30–34.

Bangle, C. 2001. The Ultimate Creativity Machine: How BMW Turns Art into Profit. *Harvard Business Review*, January–February, 5–11.

Bank, D. 1995. The Java Saga. *Wired*, December, 166–69 and 238–46.

Basel Committee Publications. 1998. No. 42. www.bis.org.

Beinhocker, E. D. 1999. Robust Adaptive Strategies. *Sloan Management Review*, Spring, 40 (3): 95–106.

Bettis, R. A., and M. A. Hitt. 1995. The New Competitive Landscape. *Strategic Management Journal* 16 (1): 7–19.

Beunza, D., and D. Stark. 2005. Resolving Identities: Successive Crises in a Trading Room after 9/11. In *Wounded City: The Social Effects of the Attack on the World Trade Center*, ed. N. Foner. New York: Russel Sage Foundation.

Boddy, D. 2002. *Managing Projects: Building and Leading the Team*. Harlow, UK: Pearson Education.

Brokaw, L. 1991. The Truth about Start-ups. *Inc*, April, 52–67.

Channel Tunnel Special Report. 1990. Handshake under the Tunnel. *ENR*, December 10, 30–31.

———. 1990. Management Turmoil Has Dogged Project from Start. *ENR*, December 10, 56–59.

Chapman, C., and S. Ward. 1997. *Project Risk Management: Processes, Techniques and Insights*. Chichester, UK: John Wiley & Sons.

Chesbrough, H. W., and S. J. Socolof. 2000. Creating New Ventures from Bell Labs Technologies. *Research Technology Management*, March–April, 13–17.

Chesbrough, H., and R. S. Rosenbloom. 2002. The Role of the Business Model in Capturing Value from Innovation: Evidence from Xerox Corporation's Technology Spinoff Companies. *Industrial and Corporate Change* 11 (3): 529–55.

Chew, W. B., D. Leonard-Barton, and R. E. Bohn. 1991. Beating Murphy's Law. *Sloan Management Review*, Spring, 5–16.

Clark, K. B., and T. Fujimoto. 1991. *Product Development Performance*. Boston: HBS Press.

Cohen, M., J. Y. Jaffray, and T. Said. 1985. Individual Behavior under Risk and Uncertainty: An Experimental Study. *Theory and Decision* 18:203–28.

Cooper R. G., S. J. Edgett, and E. J. Kleinschmidt. 1998. *Portfolio Management for New Products*. New York: Perseus Books.

Cosmides, L., and J. Tooby. 1989. Evolutionary Psychology and the Generation of Culture. *Ethology and Sociobiology* 10:51–97.

———. 1992. Cognitive Adaptations for Social Exchange. In *The Adapted Mind*, ed. J. H. Barkow, L. Cosmides, and J. Tooby. Oxford: Oxford University Press, 163–228.

Council of Standards Australia and New Zealand. 1999. *Risk Management*. Strathfield, NSW, Australia: Standards Association of Australia.

Crossan, M., M. Pina E. Cunha, D. Vera, and J. Cunha. 2005. Time and Organizational Improvisation. *Academy of Management Review* 30 (1): 129–45.

Crouhy, M., D. Galai, and R. Mark. 2000. *Risk Management*. New York: McGraw-Hill.

Cusumano, M. A., and R. W. Selby. 1995. *Microsoft Secrets: How the World's Most Powerful Software Company Creates Technology, Shapes Markets and Manages People*. New York: The Free Press.

DeMarco, T. 1997. *The Deadline. A Novel about Project Management.* New York: Dorset.

De Meyer, A., C. H. Loch, and M. T. Pich. 2002. Managing Project Uncertainty. *Sloan Management Review,* Winter, 43 (2): 60–67.

Department of Defense. 2001. *Risk Management Guide for Department of Defense Acquisitions.* Belvoir, VA: Defense Acquisition University Press.

Dohmen, F., D. Hawranek, and F. Hornig. 2004. Maut-Desaster: Eine Frage der Ehre. *Der Spiegel,* January 26.

Doz, Y. 1996. The Evolution of Cooperation in Strategic Alliances: Initial Conditions or Learning Processes? *Strategic Management Journal* 17:55–83.

Drucker, P. 1985. *Innovation and Entrepreneurship: Practice and Principles.* New York: Harper & Row.

Dvir, D., S. Lipovetky, A. Shenhar, and A. Tishler. 1998. In Search of Project Classification: A Non-Universal Approach to Project Success Factors. *Research Policy* 27:915–935.

Economist. 1989. Channel Tunnel: Under Water, Over Budget. October 7, 73–74.

———. 2004. Up, Up and Away: Tony Douglas Is Redefining How to Run Massive Construction Projects. July 24, 56.

Elmquist, S. A., E. C. Dowling, and L. A. Kipfstuhl. 2001. Cliffs and Associates Ltd. Circored Iron Plant. Paper presented at the SME International Session.

Elmquist, S. A., P. Weber, and H. Eichberger. 2001. Operational Results of the Circored Fine Ore Direct Reduction Plant in Trinidad. *Stahl und Eisen.*

Eppinger, S. D., D. E. Whitney, R. P. Smith, and D. A. Gebala. 1994. A Model-Based Method for Organizing Tasks in Product Development. *Research in Engineering Design* 6:1–13.

Eppinger, S. D., and V. Salminen. 2001. Patterns of Product Development Interactions. International Conference on Engineering Design ICED 01, Glasgow, August 21–23.

Ferreira, M. L. R., and J. H. Rogerson. 1999. The Quality Management Role of the Owner in Different Types of Construction Contracts for Process Plants. *Total Quality Management* 10 (3): 401–11.

Floricel, S., and R. Miller. 2001. Strategizing for Anticipated Risks and Turbulence in Large Scale Engineering Projects. *International Journal of Project Management* 19:445–455.

Floricel, S., and J. Lampel. 1998. Innovative Contractual Structures for Interorganizational Systems. *International Journal of Technology Management* 16 (1/2/3): 193–206.

Fox, B. L. 1993. Random Restarting versus Simulated Annealing. *Computational Mathematical Applications* 27:33–35.

Garvin, D. A. 1993. Building a Learning Organization. *Harvard Business Review* 71 (4): 78–91.

Genus, A. 1997a. Unstructuring Incompetence: Problems of Contracting, Trust, and the Development of the Channel Tunnel. *Technology Analysis and Strategic Management* 9 (4): 419–35.

———. 1997b. Managing Large-Scale Technology and Inter-Organizational Relations: The Case of the Channel Tunnel. *Research Policy* 26:169–89.

Gigerenzer, G., and K. Hug. 1993. Domain-specific Reasoning: Social Contracts, Cheating and Perspective Taking. *Cognition* 43:127–171.

Gladwell, Malcolm. 2000. *The Tipping Point.* Boston: Little, Brown and Company.

Goldratt, E. M. 1997. *Critical Chain.* Great Barrington, MA: North River Press.

Hackney, J. W. 1965. *Control and Management of Capital Projects.* New York: John Wiley & Sons.

Hamel, G., and C. K. Prahalad. 1994. *Competing for the Future.* Cambridge, MA: Harvard Business School Press.

Harmantzis, F. C. 2003. Risky Business. OR/*MS Today*, February.

Hauser, J. R. 1998. Research, Development and Engineering Metrics. *Management Science* 44 (12): 1670–689.

Herroelen, W. 2005. Project Scheduling: Theory and Practice. *Production and Operations Management*, forthcoming.

Herroelen, W., and R. Leus. 2001. On the Merits and Pitfalls of Critical Chain Scheduling. *Journal of Operations Management* 19:559–77.

Hofstede, G. 2001. *Culture's Consequences.* Thousand Oaks, CA: Sage Publications.

Hogarth, R. R. M. 2001. *Educating Intuition.* Chicago: University of Chicago Press.

Holmström, B., and P. Milgrom. 1991. Multitask Principal-Agent Analyses: Incentive Contracts, Asset Ownership and Job Design. *Journal of Law, Economics and Organizations* 7:24–52.

Iansiti, M. 1990. Microsoft Corporation: Office Business Unit. Harvard Business School Case Study 9-691-033.

Iansiti, M., and A. McCormack. 1996. Living on Internet Time: Product Development at Netscape, Yahoo, NetDynamics and Microsoft. Harvard Business School Case Study 9-697-052.

Jones, K., and T. Ohbora. 1990. Managing the Heretical Company. *McKinsey Quarterly* (3): 20–45.

Kauffman, S. A. 1993. *The Origins of Order.* Oxford, UK: Oxford University Press.

Kavadias, S., and C. H. Loch. 2004. *Project Selection under Uncertainty.* Norwell, MA, Dordrecht, Netherlands: Kluwer.

Kerzner, H. 2003. *Project Management, A Systems Approach.* 8th ed. Hoboken, NJ: John Wiley & Sons.

Kharbanda, O. P., and E. A. Stallworthy. 1984. *How to Learn from Project Disasters.* Aldershot, UK: Gower.

Kim, W. C., and R. A. Mauborgne. 1991. Implementing Global Strategies: The Role of Procedural Justice. *Strategic Management Journal* 12:125–43.

———. 1995. A Procedural Justice Model of Strategic Decision Making: Strategy Content Implications in the Multinational. *Organization Science* 6 (1): 44–61.

———— 1997. Fair Process: Managing in the Global Economy. *Harvard Business Review* 75 (4): July–August, 65–75.

Kloppenborg, T., and W. A. Opfer. 2002. The Current State of Project Management Research: Trends, Interpretations, and Predictions. *International Journal of Project Management* 33 (2): 5–19.

Kohn, A. 1993. Why Incentive Plans Cannot Work. *Harvard Business Review*, September–October, 54–63.

Kunkel, J. G. 1997. Rewarding Product Development Success. *Research Technology Management*, September–October, 29–31.

Lapré, M. A., A. S. Mukherjee, and L. N. Van Wassenhove. 2000. Behind the Learning Curve: Linking Learning Activities to Waste Reduction. *Management Science* 45 (5): 597–611.

Lasserre, P. 2003. INSEAD: One School, Two Campuses: Going to Asia. INSEAD Case Study 03/2003-5089.

Leonard-Barton, D. 1992. Core Capabilities and Core Rigidities: A Paradox in Managing New Product Development. *Strategic Management Journal* 13:111–25.

————. 1995. *Wellsprings of Knowledge.* Cambridge, MA: Harvard Business School Press.

Loch, C. H. 2003. Beauty Is in the Eye of the Beholder. In *Design Research,* B. Laurel. Cambridge, MA: MIT Press, 212–20.

————. 2005. The PCNet Project: Project Risk Management in an IT Integration Project (A and B). INSEAD Case 04/2005-5272.

Loch, C. H., S. Kavadias, and A. De Meyer. 2000. Dragonfly: Development of a UAV, INSEAD Case 03/2000-4885.

Loch, C. H., M. Yaziji, and C. Langen. 2000. The Fight for the Alpha Position: Channeling Status Competition in Organizations. *European Management Journal* 19, 16–25.

Loch, C. H., and K. Bode-Greuel. 2001. Evaluating Growth Options as Sources of Value for Pharmaceutical Research Projects. *R&D Management* 31 (2): 231–248.

Loch, C. H., C. Terwiesch, and S. Thomke. 2001. Parallel and Sequential Testing of Design Alternatives. *Management Science* 47 (5): 663–78.

Loch, C. H., and S. Tapper. 2002. Implementing a Strategy-Driven Performance Measurement System for an Applied Research Group. *Journal of Product Innovation Management* 19:185–98.

Loch, C. H., and C. Terwiesch. 2000. Product Development and Concurrent Engineering. In *Encyclopedia of Production and Manufacturing Management,* ed. P. M. Swamidass. Dordrecht, Netherlands: Kluwer Academic Publishing, 567–75.

————. 2002. Cleveland Cliffs and Lurgi GmbH: The Circored Project (A and B). INSEAD-Wharton Alliance Case and Teaching Note.

————. 2005. Rush and Be Wrong or Wait and Be Late? Seven Principles of When to Commit to Real Time Information. *Production and Operations Management* 14 (3): 331–43.

Loch, C. H., S. C. Sommer, G. Schäfer, and D. Nellessen. 2003. Will Rapid Manufacturing Bring Us the Customized Car? *Automotive World*, March 10.

Loch, C. H., and S. C. Sommer. 2004. Vol de Nuit: The Dream of the Flying Car at Lemond Automobiles. INSEAD case study.

Loch, C. H., M. E. Solt, and E. Bailey. 2005. Diagnosing and Managing Unforeseeable Uncertainty to Improve Venture Capital Returns. INSEAD working paper.

Loeb, M. 1995. Jack Welch Lets Fly on Budgets, Bonuses, and Buddy Boards. *Fortune*, May 29, 145–47.

Lynn, G. S., J. G. Morone, and A. S. Paulson. 1996. Marketing and Discontinuous Innovation: The Probe and Learn Process. *California Management Rev.* 38 (3): 8–37.

MacCormack, A. D., R. Verganti, and M. Iansiti. 2001. Developing Products on Internet Time: The Anatomy of a Flexible Development Process. *Management Science* 47 (1): 133–50.

Maclean, N. 1992. *Young Men and Fire.* Chicago: University of Chicago Press.

Maney, K. 2003. *The Maverick and His Machine: Thomas Watson Sr. and the Making of IBM.* New York: John Wiley & Sons.

McGrath, R.G. 1995. Discovery Driven Planning. *Harvard Business Review*, July–August, 44–54.

————. 2001. Exploratory Learning, Innovative Capacity, and Managerial Oversight. *Academy of Management Journal* 44 (1): 118–31.

McGrath, R.G., and I. MacMillan. 2000. *The Entrepreneurial Mindset.* Boston: Harvard Business School Press.

Meredith, J. R., and S. J. Mantel. 2003. *Project Management: A Managerial Approach.* 5th ed. New York: John Wiley & Sons.

Meulbroek, L. 2001. A Better Way to Manage Risk. *Harvard Business Review*, February, 2–3.

McDonald, D. F., and J. O. Evans III. 1998. Construction Contracts: Shifting Risk or Generating a Claim? *AACE International Transactions*, LEG.01, 1–8.

Mihm, J., and C. H. Loch. 2005. Spiraling out of Control: Problem-Solving Dynamics in Complex Distributed Engineering Projects. In *Complex Engineering Systems*, ed. D. Braha, A. Minai, and Y. Bar-Yam. New York: Perseus Books.

Miller, R., and D. L. Lessard. 2000. *The Strategic Management of Large Engineering Projects.* Boston: MIT Press.

Miner, A. S., P. Bassoff, and C. Moorman. 2001. Organizational Improvisation and Learning: A Field Study. *Administrative Science Quarterly* 46:304–37.

Mintzberg, H. 1994. *The Rise and Fall of Strategic Planning.* Hemel Hempstead, UK: Prentice Hall.

Morris, P. W. G., and G. H. Hugh. 1987. *The Anatomy of Major Projects.* Chichester, UK: John Wiley & Sons.

Nelson, R. R., and S. G. Winter. 1982. *An Evolutionary Theory of Economic Change.* Cambridge, MA: Belknap, Harvard University Press.

O'Connor, L. 1993. Tunneling under the Channel. *Mechanical Engineering,* December, 60–66.

O'Connor, G. C. and R. Veryzer. 2001. The Nature of Market Visioning for Technology Based Radical Innovation. *Journal of Product Innovation Management* 18:231–46.

Palmer, J. 1990. Deep Trouble: Financing Is Sinking for Eurotunnel. *Barron's,* September 10, 16–17 and 76–77.

Pfeffer, J. 1998. Six Dangerous Myths about Pay. *Harvard Business Review,* May–June, 109–19.

Pfeffer, J., and R. I. Sutton. 2000. *The Knowing-Doing Gap: How Smart Companies Turn Knowledge into Action.* Boston: Harvard Business School Press.

Pich, M. T., C. H. Loch, and A. De Meyer. 2002. On Uncertainty, Ambiguity and Complexity in Project Management. *Management Science* 48 (8): 1008–23.

Pillai, A. S., and K. S. Rao. 1996. Performance Monitoring in R&D Projects. *R&D Management* 26 (1): 57–65.

Pinto, J. K. 2002. Project Management 2002. *Research Technology Management,* March–April, 22–37.

Pitsis, T. S., S. R. Clegg, M. Marosszeky, and T. Rura-Polley. 2003. Constructing the Olympic Dream: A Future-Perfect Strategy of Project Management. *Organization Science* 14 (5): 574–90.

Pitt, L. F., and R. Kannemeyer. 2000. The Role of Adaptation in Microenterprise Development: A Marketing Perspective. *Journal of Developmental Entrepreneurship* 5 (2): 137–55.

PMI (Project Management Institute) Standards Committee. 1996. *A Guide to the Project Management Body of Knowledge.* Upper Darby, PA: Project Management Institute.

PMI. 2001. *Project Management Institute Fact Book.* 2nd ed. Upper Darby, PA: Project Management Institute.

Pritsker, A. A. B. 1966. *GERT: Graphical Evaluation and Review Technique.* Santa Monica: The Rand Corporation, Memorandum RM-4973-NASA.

Quinn, J. B. 1985. Managing Innovation: Controlled Chaos. *Harvard Business Review,* May–June, 73–84.

Rangan, S. R. Samii, and L. N. Van Wassenhove. 2005. Constructive Partnerships: When Alliances between Private Firms and Public Actors Can Enable Creative Strategies. *Academy of Management Review* 30, in press.

Rich, B. R. 1994. *Skunk Works: A Personal Memoir of My Years at Lockheed.* Boston: Little, Brown and Company.

Rittel, H. W. J., and M. M. Webber. 1973. Dilemmas in a General Theory of Planning. *Policy Sciences* 4:155–69.

Roberts, E. B., and A. R. Fusfeld. 1997. Critical Functions: Needed Roles in the Innovation Process. Chap. 25 in *The Human Side of Managing Technological Innovation*, ed. R. Katz. Oxford: Oxford University Press, 273–86.

Roussel P. A., K. M. Saad, and T. J. Erickson. 1991. *Third Generation R&D*. Boston: Harvard Business School Press.

Sabbagh, K. 1996. *Twenty-First Century Jet*. New York: Scribner.

Sahlman, W. A. 1990. The Structure and Governance of Venture-Capital Organizations. *Journal of Financial Economics* 27:473–521.

Schein, E. H. 1985. *Organizational Culture and Leadership*. San Francisco: Jossey-Bass.

Schoemaker, P. J. H. 2002. *Profiting from Uncertainty: Strategies for Succeeding No Matter What the Future Brings*. New York, NY: Free Press.

Schrader, S., W. M. Riggs, and R. P. Smith. 1993. Choice over Uncertainty and Ambiguity in Technical Problem Solving. *Journal of Engineering and Technology Management* 10:73–99.

Sengupta, K., and T. K. Abdel-Hamid. 1993. Alternative Conceptions of Feedback in Dynamic Decision Environments: An Experimental Investigation. *Management Science* 39 (4): 411–28.

Shenhar, A. J. 1998. From Theory to Practice: Toward a Typology of Project Management Styles. *IEEE Transactions on Project Management* 45 (1): 33–48.

———. 2001. One Size Does Not Fit All Projects. *Management Science* 47 (3): 394–414.

Shenhar, A. J., and D. Dvir. 1996. Toward a Typology Theory of Project Management. *Research Policy* 25:607–32.

Simon, H. A. 1951. A Formal Theory of the Employment Relationship. *Econometrica* 19(3): 293–305.

———. 1955. A Behavioral Model of Rational Choice. *Quarterly Journal of Economics* 69:99–118.

———. 1969. *The Science of the Artificial*. Boston, MA: MIT Press.

Simons, R. L. 1999. A Note on Identifying Strategic Risk. Harvard Business School Case 9-199-031, revised.

Sinclair-Desgagné, B. 1999. How to Restore High-Powered Incentives in Multitask Agencies. *Journal of Law, Economics and Organization* 15(2): 418–33.

Smith, R. P. 1997. The Historical Roots of Concurrent Engineering Fundamentals. *IEEE Transactions on Engineering Management* 44 (1): 67–78.

Smith, R., and I. Walter. 1989. Eurotunnel – Background. INSEAD-New York University Case Study 08/95-2492.

Smith, P. G., and G. M. Merritt. 2002. *Proactive Risk Management*. New York: Productivity Press.

Sobek, D. K. II, A. C. Ward, and J. K. Liker. 1999. Toyota's Principles of Set-Based Concurrent Engineering. *Sloan Management Review* 40:67–83.

Sobiezczansky-Sobiesky, J., J. S. Agde, and R. R. Sandusky. 1998. Bi-level Integrated System Synthesis (BLISS). *7th AIAA Symposium Multidisciplinary Analytical Optimization Collected Technical Papers*, Pt. 3, A98-39701, St. Louis, MO, 10–31.

Soderberg, L. G. 1989. Facing up to the Engineering Gap. *The McKinsey Quarterly*, Spring, 3–23.

Sommer, S. C. 2004. Managing Projects Under Unforeseeable Uncertainty and Complexity. Unpublished Ph.D. dissertation, INSEAD, June.

Sommer, S. C., and C. H. Loch. 2004. Selectionism and Learning in Projects with Complexity and Unforeseeable Uncertainty. *Management Science* 50 (10): 1334–47.

———. 2005. Incentive Contracts in Projects with Unforeseeable Uncertainty. INSEAD and Purdue University working paper.

———. 2005b. Incentive Contracts for Parallel Trials. INSEAD and Purdue University working paper.

Sosa, M., S. D. Eppinger, and C. Rowles. 2005. The Misalignment of Product Architecture and Organizational Structure in Complex Product Development. *Management Science* 51, forthcoming.

Stalk, G. Jr., A. M. Webber. 1993. Japan's Dark Side of Time. *Harvard Business Review* 71 (4): 93–102.

Steward, D. V. 1981. *Systems Analysis and Management: Structure, Strategy and Design*. New York: Petrocelli Books.

Stewart, T. A. 1995. Planning a Career in a World without Managers. *Fortune*, March 20, 40–45.

Stinchcombe, A. L., and C. A. Heimer. 1985. *Organization Theory and Project Management*. Oslo: Norwegian University Press.

Tampoe, M., and L. Thurloway. 1993. Project Management: The Use and Abuse of Techniques and Teams. *International Journal of Project Management* 11 (4): 245–50.

Tatikonda, M. V., and S. R. Rosenthal. 2000. Technology Novelty, Project Complexity and Product Development Project Execution Success. *IEEE Transactions on Engineering Management* 47 (1): 74–87.

Teece, D. J., G. P. Pisano, and A. Shuen. 1997. Dynamic Capabilities and Strategic Management. *Strategic Management Journal* 18 (7): 509–33.

Terwiesch, C., C. H. Loch, and A. De Meyer. 2002. Exchanging Preliminary Information in Concurrent Engineering: Alternative Coordination Strategies. *Organization Science* 13 (4): 402–19.

Terwiesch, C., and C. H. Loch. 2004. Collaborative Prototyping and the Pricing of Customized Products. *Management Science* 50 (2): 145–58.

Thomke, S. 2001. Enlightened Experimentation: The New Imperative for Innovation. *Harvard Business Review*, February, 74 (1): 66–75.

———. 2003. *Experimentation Matters.* Cambridge, MA: Harvard Business School Press.

Thomke, S., and D. Reinertsen. 1998. Agile Product Development: Managing Development Flexibility in Uncertain Environments. *California Management Review* 41 (1): 8–30.

Trivers, R. L. 1971. The Evolution of Reciprocal Altruism. *Quarterly Review of Biology* 46:35–57.

Ulrich, A. 2003. Verkehr: Stück aus dem Tollhaus. *Der Spiegel*, September 15.

Ulrich, K., and S. D. Eppinger. 2004. *Product Design and Development*. 3rd ed. New York: McGraw-Hill.

Urda, J., and C. H. Loch. 2005. Social Appraisals: How the Social Environment Triggers Emotions. INSEAD working paper.

Van de Ven, A. H. 1986. Central Problems in the Management of Innovation. *Management Science* 32 (5): 590–607.

Van de Ven, A. H., D. E. Polley, R. Garud, and S. Venkataraman. 1999. *The Innovation Journey*. Oxford: Oxford University Press.

Von Bitter, R., R. Husain, P. Weber, and H. Eichberger. 1999. Experiences with Two New Fine Ore Reduction Processes, Circored and Circofer. Paper presented at the METEC Conference, Düsseldorf, Germany.

VF. 1989. Managing a Mega Project. *Civil Engineering*, June, 44–47.

Von Branconi, C., and C. H. Loch. 2004. Contracting for Major Projects: Eight Business Levers for Top Management. *International Journal of Project Management* 22 (2): 119–30.

Ward, S., and C. Chapman. 1994. Choosing Contractor Payment Terms. *International Journal of Project Management* 12 (4): 216–21.

Ward, A., J. K. Liker, J. J. Crisitiano, and D. K. Sobek II. 1995. The Second Toyota Paradox: How Delaying Decisions Can Make Better Cars Faster. *Sloan Management Review*, Spring, 43–61.

Watts, R. M. 2001. Commercializing Discontinuous Innovations. *Research Technology Management*, November–December, 26–31.

Weick, K. E. 1993. The Collapse of Sensemaking in Organizations: The Mann Gulch Disaster. *Administrative Science Quarterly* 38(4): 628–52.

———. 1995. *Sensemaking in Organizations*. Thousand Oaks, CA: Sage Publications.

Weick, K. E., and K. M. Sutcliffe. 2001. *Managing the Unexpected*. San Francisco: Jossey-Bass.

West, J. 2000. Institutions, Information Processing, and Organization Structure in Research and Development: Evidence from the Semiconductor Industry. Research Policy 29 (3): 349–73.

Wheelwright, S. C., and K. B. Clark. 1992. *Revolutionizing Product Development*. New York: The Free Press.

Wideman, R. M. 1992. *Project and Program Risk Management*. Newton Square, PA: Project Management Institute.

Williams, T. M. 1999. The Need for New Paradigms for Complex Projects. *International Journal of Project Management* 17 (5): 269–73.

———. 2002. *Modelling Complex Projects.* Chichester, UK: John Wiley & Sons.

Zack, J. G. Jr. 1996. Risk Sharing: Good Concept, Bad Name. *Cost Engineering* 38 (7): 26–33.

Index